Object-based Programming with Go

Christian Maurer

Object-based Programming with Go

Christian Maurer
Berlin, Germany

ISBN 978-3-658-44703-8   ISBN 978-3-658-44704-5   (eBook)
https://doi.org/10.1007/978-3-658-44704-5

This book is a translation of the original German edition "Objektbasierte Programmierung mit Go," 2nd edition, by Christian Maurer, published by Springer Fachmedien Wiesbaden GmbH in 2023. The translation was done with the help of an artificial intelligence machine translation tool. A subsequent human revision was done primarily in terms of content, so that the book will read stylistically differently from a conventional translation. Springer Nature works continuously to further the development of tools for the production of books and on the related technologies to support the authors.

© The Editor(s) (if applicable) and The Author(s), under exclusive license to Springer Fachmedien Wiesbaden GmbH, part of Springer Nature 2025

This work is subject to copyright. All rights are solely and exclusively licensed by the Publisher, whether the whole or part of the material is concerned, specifically the rights of translation, reprinting, reuse of illustrations, recitation, broadcasting, reproduction on microfilms or in any other physical way, and transmission or information storage and retrieval, electronic adaptation, computer software, or by similar or dissimilar methodology now known or hereafter developed.
The use of general descriptive names, registered names, trademarks, service marks, etc. in this publication does not imply, even in the absence of a specific statement, that such names are exempt from the relevant protective laws and regulations and therefore free for general use.
The publisher, the authors and the editors are safe to assume that the advice and information in this book are believed to be true and accurate at the date of publication. Neither the publisher nor the authors or the editors give a warranty, expressed or implied, with respect to the material contained herein or for any errors or omissions that may have been made. The publisher remains neutral with regard to jurisdictional claims in published maps and institutional affiliations.

Planung/Lektorat: Leonardo Milla
This Springer imprint is published by the registered company Springer Fachmedien Wiesbaden GmbH, part of Springer Nature.
The registered company address is: Abraham-Lincoln-Str. 46, 65189 Wiesbaden, Germany

If disposing of this product, please recycle the paper.

Dedicated to Professor David L. Parnas

Preface

Object-oriented languages
A side effect of the application of information hiding
is the creation of new objects that store data.
...FORTRAN ...Pascal ...Simula ...Smalltalk ...
More recent languages have added new types of features
(known as inheritance)
designed to make it possible
to share representations between objects.
Often, these features are misused
and result in a violation of information hiding
and programs that are hard to change.

The most negative effect of the development of O-O languages has been
to distract programmers from design principles.
Many seem to believe that
if they write their program in an O-O language,
they will write O-O programs.
Nothing could be further from the truth.

Component-Oriented Design
The old problems and dreams are still with us.
Only the words are new.

Abstract Data types
...Being able to use variables of these new,
user-defined, abstract data types
in exactly the way as we use variables of built-in data types
is obviously a good idea.
Unfortunately, I have never seen a language that achieved this.

David L. Parnas
In: The Secret History of Information Hiding, Software Pioneers, Springer 2002
This book consists of two parts:

- The implementation of object-based programming with Go,
  - a presentation of the basics of object-based development,
  - an introduction to essential aspects of Go and
  - the introduction of the microuniverse μU with the presentation of various classic algorithms,
- the documentations of teaching projects from computer science teacher training courses at the Institute for Computer Science of the Free University of Berlin and some of my program systems based on them:
  - the robots,
  - the appointment calendar,
  - the game of life,
  - the Go register machine,
  - the electronic stylus,
  - the single-address machine Mini,
  - the management of a book inventory,
  - the Inferno, a management of almost any data sets,
  - the Lindenmayer systems,
  - the operation of train stations,
  - the representation of figures in space, and
  - the Berlin's U- and S-Bahn networks.

I would like to express my sincere thanks to Mr. Leonardo Milla and Mrs. Juliane Wagner from Springer-Verlag. They very kindly supported the idea to translate the second edition of my book "Objektbasierte Programmierung with Go".

All source codes are available on the book's page on the World Wide Web:
`https://maurer-berlin.eu/obpbook`.

Berlin,
August 2024

Christian Maurer

Contents

Part I The Realisation of Object-based Programming with Go

1 Basics of Object-Oriented Development 3
   1.1   The Program Life Cycle. 3
         1.1.1   System Analysis. 4
         1.1.2   System Architecture. 6
         1.1.3   User Manual. 10
         1.1.4   Construction. 11
   1.2   Advantages of an Object-Based System Architecture 13
         1.2.1   On Specification. 13
         1.2.2   On Implementation. 14
         1.2.3   On Implementation. 14

2 Aspects of Go. 17
   2.1   About the Installation of Go 18
   2.2   Packages in Go. 19
         2.2.1   Program Packages 19
         2.2.2   Packages as Interfaces Only. 22
         2.2.3   Nesting of Packages. 22
         2.2.4   Initialization of Packages. 22
         2.2.5   Variables of Concrete Data Types 23
         2.2.6   References and Parameters 25
   2.3   Variables of Abstract Data Types = Objects 26
   2.4   Value Versus Reference Semantics 28
         2.4.1   Assignments, Creation of Copies. 28
         2.4.2   Equality Check and Size Comparison 29
         2.4.3   Serialization. 29

3 The Microuniverse 31
   3.1   Installation of the Microuniverse. 31
         3.1.1   Prerequisites. 32

		3.1.2	License Terms	34
		3.1.3	Naming in the Microuniverse	34
	3.2	The Constructor New		35
	3.3	The Object Package		36
		3.3.1	Equaler	36
		3.3.2	Comparer	38
		3.3.3	Clearer	38
		3.3.4	Coder	39
		3.3.5	The Interface of the Package Obj.	41
		3.3.6	Stringer	43
		3.3.7	Formatter	43
		3.3.8	Valuator	43
	3.4	Input and Output		44
		3.4.1	Packages for the Screen	44
		3.4.2	Screen	48
		3.4.3	Keyboard	59
		3.4.4	Editor	62
		3.4.5	Input/Output Fields	63
		3.4.6	Error Messages and Hints	65
		3.4.7	Printer	66
		3.4.8	Selections	67
		3.4.9	Menues	68
	3.5	Collections of Objects		69
		3.5.1	Collector	69
		3.5.2	Seeker	72
		3.5.3	Predicator	72
		3.5.4	Sequences	74
		3.5.5	Stacks	78
		3.5.6	Buffers (Queues)	78
		3.5.7	Priority Queues	80
		3.5.8	Sets	83
		3.5.9	Persistent Sequences (Sequential Files)	97
		3.5.10	Persistent Index Sets	99
		3.5.11	Graphs	103
	3.6	Additional Data Types from the Microuniverse		115

Part II The Projects

4 General ... 119
    4.1 Teaching Projects ... 120
        4.1.1 System Analysis ... 121

		4.1.2	System Architecture	123
		4.1.3	User Manual	123
5	**Robi**			**125**
	5.1	System Analysis		126
	5.2	The Robi Language		127
	5.3	System Architecture		128
	5.4	User Manual		128
		5.4.1	The Robi Editor	128
		5.4.2	The Robi Protocol	129
		5.4.3	Robi-Sokoban	129
		5.4.4	Robot Race	129
		5.4.5	General Procedure	129
	5.5	Construction		130
	5.6	Exercises		133
		5.6.1	Sample Solutions	134
6	**The Appointment Calendar**		**139**
	6.1	System Analysis		139
		6.1.1	Calendar Pages	140
		6.1.2	Day Attributes	140
		6.1.3	Sequences of Appointments and Appointments	140
		6.1.4	Annual Calendar	142
		6.1.5	Monthly and Weekly Calendars	142
		6.1.6	Appointment Calendar	143
		6.1.7	Search for Appointments	144
	6.2	System Architecture		144
		6.2.1	The Objects of the System	144
		6.2.2	Component Hierarchy	144
	6.3	User Manual		145
		6.3.1	Formats	146
	6.4	Calendar Pages		147
	6.5	Weekly Calendar		147
		6.5.1	Monthly Calendar	148
	6.6	Annual Calendar		148
		6.6.1	Search and Search Results	148
		6.6.2	System Operation	149
	6.7	Construction		151
		6.7.1	Term Attributes	151
		6.7.2	Keywords	152
		6.7.3	Appointments	152
		6.7.4	Appointment Sequences	152

		6.7.5	Persistent Sets of Calendar Data	153
		6.7.6	Day Attributes	153
	6.8	Calendar Pages		154
		6.8.1	Appointment Calendars	154

7 Life ... 157

	7.1	System Analysis		157
		7.1.1	The Game of Life	158
		7.1.2	The Ecosystem of Foxes, Rabbits, and Plants	158
		7.1.3	The Objects of the System	159
		7.1.4	Component Hierarchy	159
	7.2	User Manual		159
		7.2.1	Program Operation	160
	7.3	Construction		161
		7.3.1	Specifications	161
		7.3.2	Implementations	163

8 The Go Register Machine ... 165

	8.1	System Analysis		166
		8.1.1	Components of a Register Machine	166
		8.1.2	Basics of the Register Machine Programming Language	167
		8.1.3	System Architecture	167
		8.1.4	Registers	168
		8.1.5	Register Machine Programs	168
		8.1.6	Instructions	169
		8.1.7	Test Programs	172
		8.1.8	Functions	173
	8.2	User Manual		175
		8.2.1	Examples	176
		8.2.2	Recursion	178
	8.3	Construction		184
	8.4	Exercises		186

9 The Electronic Stylus ... 189

	9.1	System Analysis		189
		9.1.1	The Figures of the Electronic Stylus	191
		9.1.2	The Operations on the eBoards	193
		9.1.3	Program Start	194
		9.1.4	Program Start	194
		9.1.5	Creation of New Figures	195
		9.1.6	Modification of Figures	197
		9.1.7	Deleting of Figures	197
		9.1.8	Marking Figures	197

	9.1.9	Loading and Saving	198
	9.1.10	Printing	198
	9.1.11	Brief Help	198
	9.1.12	System Architecture	198
9.2	Construction		198

10 Mini 201

10.1	System Analysis		202
	10.1.1	Processor	202
	10.1.2	Data Storage	202
	10.1.3	Program Lines	203
	10.1.4	Execution of a Mini Program	203
	10.1.5	Instructions	204
	10.1.6	Example	207
	10.1.7	The Objects of the System	208
	10.1.8	Component Hierarchy	208
10.2	User Manual		209
10.2.1	Instructions for Working with Mini		210
10.3	Construction		210
10.4	Exercises		211

11 Books 213

11.1	System Analysis		213
11.2	System Architecture		214
11.3	The Objects of the System		214
11.4	Component Hierarchy		215
11.5	User Manual		215
11.6	Construction		216
	11.6.1	Areas	216
	11.6.2	Natural Numbers	216
	11.6.3	Strings	217
	11.6.4	Book	218
	11.6.5	Books	219
	11.6.6	The Program for Managing the Book Inventory	219

12 Inferno 221

12.1	System Analysis		221
	12.1.1	Masks	222
	12.1.2	Molecules	222
	12.1.3	Structure of the Molecules	222
	12.1.4	Atoms	223
12.2	System Architecture		223
	12.2.1	The Objects of the System	223

		12.2.2	Component Hierarchy	224
		12.2.3	The Objects of the System	224
	12.3	User Manual		225
		12.3.1	Construction of an Inferno Program	225
		12.3.2	System Operation	228
		12.3.3	Construction	229
		12.3.4	Molecules	229
		12.3.5	Structure	232
		12.3.6	Atoms	233
13	**Lindenmayer Systems**			**237**
	13.1	System Analysis		237
		13.1.1	Alphabets, Languages, and Grammars	238
		13.1.2	Relationship Between Grammars and Languages	238
	13.2	The Grammars of Lindenmayer Systems		239
	13.3	Graphical Interpretation of L-Systems		239
		13.3.1	The Koch Islands	240
		13.3.2	The Islands and Lakes	240
		13.3.3	The Pavement	242
		13.3.4	Space-Filling Curves	244
		13.3.5	Extensions of the Alphabet of L-Systems	247
		13.3.6	Three-Dimensional L-Systems	251
	13.4	System Architecture		258
		13.4.1	The Objects of the System	259
		13.4.2	Component Hierarchy	260
	13.5	User Manual		261
		13.5.1	Creation of an L-System	261
		13.5.2	System Operation	261
	13.6	Construction		261
		13.6.1	Specification of the Library Packages	261
		13.6.2	Implementation of the Packages	264
14	**Rail**			**267**
	14.1	System Analysis		269
		14.1.1	Basic Concepts of Railway Technology	269
		14.1.2	Sources	270
		14.1.3	Track Diagram Display	273
		14.1.4	Driving Orders	273
		14.1.5	Representation of Train Journeys	274
	14.2	System Architecture		274
		14.2.1	The Objects of the System	274
		14.2.2	Component Hierarchy	275

14.3	User Manual		276
	14.3.1	Screen Design	276
	14.3.2	The Track Diagram Control Panel on the Screen	277
	14.3.3	The Net of the Stations	278
	14.3.4	The Network of Stations	278
	14.3.5	System Operation	280
14.4	Construction		281
	14.4.1	Main Program	282
	14.4.2	Network	282
	14.4.3	Stations	284
	14.4.4	Routes	284
	14.4.5	Blocks	285
	14.4.6	Cells	289
	14.4.7	Signals	291
	14.4.8	Aid Packages	291
	14.4.9	Other Packages	293

15 Figures in Space ... 295
    15.1 System Analysis ... 295
        15.1.1 System Architecture ... 296
        15.1.2 The Objects of the System ... 296
    15.2 Component Hierarchy ... 296
    15.3 User Manual ... 297
    15.4 Construction ... 297
        15.4.1 Specifications ... 297
        15.4.2 Implementations ... 298
        15.4.3 Examples ... 299
        15.4.4 Examples of Conic Sections ... 301

16 Berlin's U- and S-Bahn ... 305
    16.1 System Analysis ... 305
    16.2 System Architecture ... 306
    16.3 The Objects of the System ... 306
    16.4 Component Hierarchy ... 306
    16.5 User Manual ... 307
    16.6 Construction ... 307
        16.6.1 Specifications ... 307
        16.6.2 Implementation ... 310

Index ... 313

List of Figures

Fig. 1.1	The order	5
Fig. 1.2	How the system analysts understood the order	5
Fig. 1.3	Reduction of the system analysis, because the clients are stingy	5
Fig. 1.4	System Architecture	6
Fig. 1.5	Construction	11
Fig. 1.6	What *actually* was meant ...	13
Fig. 3.1	1214 N is to be inserted before 1214 A	76
Fig. 3.2	1214 N is inserted before vor 1214 A	77
Fig. 3.3	1214 A is to be removed	77
Fig. 3.4	1214 A is deleted	77
Fig. 3.5	Heap with 12 numbers	82
Fig. 3.6	Heap with 13 numbers	82
Fig. 3.7	AVL-Baum mit zwei Zahlen	85
Fig. 3.8	Tree with three numbers	85
Fig. 3.9	AVL tree with mit three numbers	85
Fig. 3.10	AVL tree with 11 numbers	85
Fig. 3.11	AVL tree with 12 numbers	86
Fig. 3.12	Baum mit 12 Zahlen	87
Fig. 3.13	AVL tree with 12 numbers	87
Fig. 3.14	AVL tree with 12 numbers	89
Fig. 3.15	Baum mit 11 Zahlen	89
Fig. 3.16	AVL-Baum mit 11 Zahlen	89
Fig. 3.17	Example of a graph	104
Fig. 5.1	The city from the first exercise	135
Fig. 5.2	The maze from the second exercise with 13 blocks	136
Fig. 6.1	The annual calendar with entered vacation times	142
Fig. 6.2	System architecture of the appointment calendar	145
Fig. 7.1	Component hierarchy of the Game of Life	159
Fig. 7.2	The Game of Life: the gun	161
Fig. 7.3	An ecosystem	162

Fig. 10.1	System architecture of Mini	209
Fig. 11.1	System architecture of the management of the book inventory	215
Fig. 11.2	The screen mask	215
Fig. 12.1	System architecture of Inferno	224
Fig. 12.2	Window of the example	225
Fig. 13.1	Koch Island: start	240
Fig. 13.2	Koch Island after 1 application step	241
Fig. 13.3	Koch Island after 2 application steps	241
Fig. 13.4	Koch Island after 3 application steps	241
Fig. 13.5	Koch Island after 4 application steps	242
Fig. 13.6	Islands and lakes	243
Fig. 13.7	Pavement after two application steps	243
Fig. 13.8	Pavement after five application steps	244
Fig. 13.9	Hilbert curve after two application steps	245
Fig. 13.10	Hilbert curve after four application steps	245
Fig. 13.11	Hilbert curve after seven application steps	246
Fig. 13.12	Peano-curv after two application steps	246
Fig. 13.13	Peano-curv after four application steps	247
Fig. 13.14	Barrel curve after two application steps	247
Fig. 13.15	Barrel curve after four application steps	248
Fig. 13.16	Sierpinski curve after two application steps	248
Fig. 13.17	Sierpinski curve after six application steps	249
Fig. 13.18	Two herbs	250
Fig. 13.19	A bush	251
Fig. 13.20	Two herbs	252
Fig. 13.21	Another herb	253
Fig. 13.22	Three-dimensional Hilbert curve	254
Fig. 13.23	Another view of the three-dimensional Hilbert curve	254
Fig. 13.24	A three-dimensional bush	255
Fig. 13.25	Another view of the three-dimensional bush	255
Fig. 13.26	Simple three-dimensional tree	256
Fig. 13.27	Three-dimensional tree	257
Fig. 13.28	Other view of the three-dimensional tree	257
Fig. 13.29	Simple three-dimensional flower	258
Fig. 13.30	Three-dimensional grass plant	259
Fig. 13.31	Three-dimensional grass plant from above	259
Fig. 13.32	Three-dimensional fantasy plant	260
Fig. 13.33	System architecture of the L-System	260
Fig. 14.1	Architecture of `Rail`	275
Fig. 14.2	Track cells	276
Fig. 14.3	Track bends	276
Fig. 14.4	Switches branched in the direction of the kilometerage	277

List of Figures

Fig. 14.5	Switches branched against the direction of the kilometerage	277
Fig. 14.6	Double crossing switches	277
Fig. 14.7	Buffer stops	277
Fig. 14.8	The net of the six stations	278
Fig. 14.9	Track diagram of Bahnheim	278
Fig. 14.10	Track diagram of Bahnhausen	279
Fig. 14.11	Track diagram of Bahnstadt	279
Fig. 14.12	Track diagram of Eisenstadt	280
Fig. 14.13	Track diagram of Eisenhausen	280
Fig. 15.1	Component hierarchy of the spatial figures	296
Fig. 15.2	Several figures	300
Fig. 15.3	Another view of the several figures	300
Fig. 15.4	Sphere, tori, and cylinder	301
Fig. 15.5	Section of a cone with a plane	302
Fig. 15.6	Another view of this section	302
Fig. 15.7	Section of a double cone with a plane parallel to the cone axis	303
Fig. 15.8	The hyperbola	303
Fig. 16.1	Architecture of BUS	307
Fig. 16.2	Extract from the U- and S-Bahn-Net in Berlin	308

Part I
The Realization of Object-Based Programming with Go

Basics of Object-Oriented Development

> *Entia non sunt multiplicanda praeter necessitatem;*
> *frustra fit per plura, quod fieri potest per pauceriora.*
>
> Johannes Clauberg (1622–1665)
> attributed to William of Ockham (1287–1347)
>
> *Entities should not be multiplied beyond necessity;*
> *it is futile to do with more what can be done with fewer.*
> Occam's razor

Abstract

This chapter presents a brief characterization of a program life cycle reduced to its essential core. The task of system analysis is to isolate the objects that occur in a system. These objects provide the components of the system architecture and thus a stringent concept for construction.

The central guiding idea that we are pursuing here is that all constructions in the context of the development of program(system)s are based primarily on the *systematic development of abstract data types*. The principles presented are based on what Parnas taught us in [5] in the early 70s (see also, e.g., [1] and [2]).

They are universally valid insofar as they are largely independent of specific programming paradigms. (The restriction "largely" is justified by the fact that the *state concept* of imperative programming clearly shines through at many points, which makes no sense in the *declarative* paradigm).

1.1 The Program Life Cycle

The core of all models of a *software life cycle* is the following phases:

- *System analysis*.
- *System architecture*.

- *User manual.*
- *Construction.*

The maintainability of systems is determined by the following *basic principles of analysis, planning, design, and implementation*:

- the detailed examination of all *factual backgrounds* of the task at hand,
- a *decomposition* into components and the description of their mutual dependencies as well as
- the complete and consistent definition of the *external behaviour* of the system,
- the elegant and comprehensible *description* and *construction* of the identified *components*.

Lack of consideration for these principles results in error-prone, uncontrollable, and risky systems, whose

- intended behaviour;
- adaptability to other machines, operating systems, development environments, or programming languages;
- developability and maintainability in case of changes or updates to the requirements

cannot be fundamentally guaranteed due to their inherent instability against small changes and whose *parts* are also *not usable* for solving other problems.

Conversely, this characterizes some minimal requirements for the development of programs that were articulated in the "software crisis" around 1970, which led to the *software engineering* becoming an independent field of computer science.

Every phase model ultimately assumes a rigid concept and does not sufficiently take into account the dialectical interplay of the phases with each other.

1.1.1 System Analysis

For every project to construct an IT system, investigations into the functional processes and data flows in the system are necessary to specify the order, especially about which parts of the system to be automated can be handled by computers. They form the necessary prerequisites for determining the performance of the IT system (see Fig. 1.1) and thus for formulating the order.

In addition, there is the dialogue between clients and system analysts about details of the system's purpose, which ultimately forms the basis for the system analysis (see Fig. 1.2).

Fig. 1.1 The order

Fig. 1.2 How the system analysts understood the order

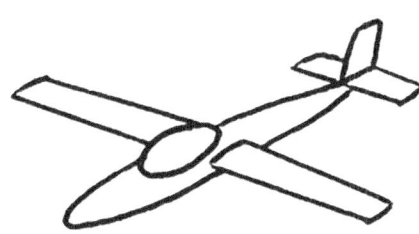

In the course of a deeper engagement with the subject matter, the repercussions of computer use must also be included: The structure of the system under consideration may change due to the switch to automatic data processing.

In connection with such considerations, sensitivity to the risks of blind trust in IT systems, which are based on *human-written* (sic!) *texts*—the source texts of programs—which are neither "tangible" nor objectively "measurable", but *pure mental constructions*, also grows.

Working on the system analysis can lead to the realization that not necessarily *every* aspect of the system can be automated, because the realization of some interesting idea within the planned cost-benefit ratio proves to be too expensive (see Fig. 1.3).

The detailed considerations in the system analysis provide a natural entry into the design work, because the *objects* recognized in the system and their *structure* can be derived from the factual analysis.

What will be shown in the following chapter is already assured here:

▶   Along these objects, the answer to the question of how to break down the system into manageable parts arises entirely on its own.

Fig. 1.3 Reduction of the system analysis, because the clients are stingy

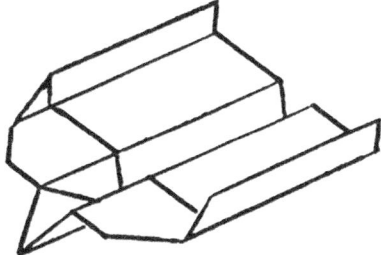

Fig. 1.4 System Architecture

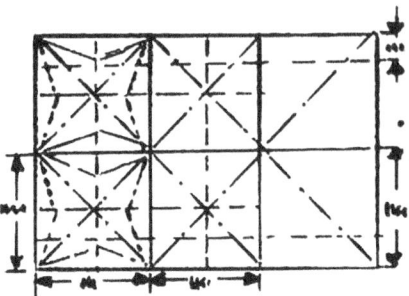

1.1.2 System Architecture

Following the system analysis is the work on the *design of the system architecture* (see Fig. 1.4) with the aim of breaking down the overall system into components and their mutual dependency.

The guiding principle here is the question of what the individual parts of the system are and how they are connected. The main concern in this phase is the *reduction of the system's complexity* to a manageable level, which ideally results from the findings of the system analysis.

The following postulates serve to obtain meaningful criteria for a breakdown into components:

- a strong *internal connection* of each individual component and
- an *understandability*, *constructability*, *testability* and *maintainability* that is largely independent of the other components.

To fulfill these requirements, each component must be split into two parts:

- the *specification*, which is a list of all its services and the exact description of the prerequisites and effects for each individual service and
- the *implementation* of these services according to the specification, in which a design decision is made taking into account the requirement profile for the system behaviour (such as optimizing runtime behaviour or memory usage).

A consequence of distinguishing these two parts is the demand for a *strict separation* of specification and implementation in different text files with a number of advantages:

- The specifications can be protected against subsequent changes by implementers (a measure that represents a protection mechanism against typical difficulties in the construction of larger systems).

- People who use the components to develop their own components are not overwhelmed by the fact that their work requires knowledge of the implementations of the components used; they only need to know their specification.
- The implementation of different alternatives by different people is possible.

As a conclusion from these considerations, it follows that programming languages must be used for the development of a system that can realize this concept.

The strict separation of the two parts ensures that clients of a component do not make implicit assumptions about its behaviour that they have from the knowledge of implementation details. Only in this way are the internal data of the component safe from uncontrolled access at the "*interface*" (= the specification) that can change its behaviour and thereby generate side effects that can have completely unpredictable effects on the system.

(How could a motor vehicle be developed, for example, if the construction of the body depended on technical details of the cylinder head cover or the anti-lock braking system, or even tried to influence these details?)

1.1.2.1 Characterization of the Component Concept

The following specifies the general requirements for the components of a decomposition of the previous section:

Necessary conditions for a clean component concept are

- the *simplicity* of the specification of the components and
- the *context independence* of their implementations. .

The *simplicity* of the specification of a component includes

- *precise colloquial formulations*, possibly functional specifications (i.e., in a functional programming language), algebraic specifications, or in formal specification languages;
- *minimality* of its scope by providing a coherent, non-decomposable problem circle;
- simultaneously *maximality* of its scope with the aim of usability for other purposes than originally planned, but still *openness* for extensions of its scope;
- *independence* from the specifications of other components except those on which they build by "extension of the specification";
- the reduction of data transports to the minimum possible extent, both within the component and between it and the components it uses;
- the rigid *avoidance of revealing any implementation details*.

The following points can be assigned to the *context independence of the implementation* of a component:

- *limitation* to the completion of the task given by the specification, which is characterized by a strong internal (logical and factual) binding, thus renouncing the construction of system parts that do not directly arise from the given specification;
- limiting the number of used components to the minimum necessary for the fulfillment of the specification task, possibly by outsourcing separable parts to separate components;
- keeping open *alternative implementations*, e.g., from the efficiency point of view of the intended purpose;
- the independent selection of such data structures and algorithms that are adapted to the requirement profile;
- *coupling* to the used components only via their specification, thus without any knowledge of their internal data or processes;
- *testability*, i.e., verifiability of their proper function according to specification;
- *maintainability*, i.e., localizability and correctability of errors and adaptability to other conditions of use.

1.1.2.2 Object-Based Decompositions

As early as 1972, Parnas formulated a—at the time still unconventional—decomposition criterion in [6]:
instead of decomposing a system according to its process steps, each component should implement a design decision (for which there are usually alternatives).

This claim is fulfilled by a *system architecture* that is oriented towards the *objects* of the system under consideration oriented:

▶ The components of a system are defined by its objects and the properties and operations that characterize or process them.

Such an *object-based decomposition* can not only be derived in a very natural way from system analysis, but also provides, as will be shown below, a stringent approach to the system architecture of a system: From it result *sufficient* conditions for a definable component concept in the sense of the previous section, i.e., it fully meets all the demands mentioned.

A *process-oriented* decomposition, on the other hand, does *not at all* correspond to the path outlined by Parnas through the program life cycle.

The approach to an object-based design methodology is the task of modelling the real objects that are manipulated in the system at an appropriate level of detail from the system analysis and working out the accesses to their models from their performance spectrum. Such a structural analysis leads retrospectively to a deeper understanding of the underlying real system and therefore also to the demand for accuracy of the formulations in the requirement definition. This leads to the determination

- of both the specifications of individual components
- and the interrelationships of various components in the system architecture.

The hierarchy of the thus found—initially unordered—set of the system's components results from the uncovering of the dependencies between the considered objects: A component that defines new objects by structuring given objects uses exactly those components that in turn define the objects to be combined.

1.1.2.3 Abstract Datatypes

The demands of the previous section for strong internal coherence and a high degree of independence from other components are immediately met when a component

- either handles a *class of objects of the same type* or
- —in exceptional cases—manages access to a *single* object.

In programming languages that allow a separation of specification and implementation, the associated specification

- defines an *abstract data type*, i.e., a class of objects with their access operations, which is *abstract* in the sense that its representation in the specification may be described in a comment-like manner, but is not syntactically visible or
- the access operations to an abstract data object (which is only managed in the implementation, so it is not explicitly provided).

The second case can be considered a special case of the first, as it concerns *one* instance of an abstract data type and the access functions to it.

An exception is those components of the lowest layer that connect the system to the services of the operating system; individual peripheral components of the computer (such as screen, keyboard, mouse, or printer) are usually only needed in one form and are therefore usually modelled as single data objects.

Further component types are superfluous, because even algorithmically emphasized system parts gain significantly in clarity structurally and fit cleanly into the hierarchy of the objects used in the system when they are recognized as access operations on certain data types to be worked out. Thus, a decomposition can largely rely on the first-mentioned case.

The implementation of a data type and the accesses to it are again composed of data types and the accesses to them, which are specified in other such components.

The decomposition of a system into components has reached an end *exactly* when an end is reached, when only atomic data types, i.e., components of the used programming language, are used in the implementation, from which all data types are ultimately composed.

Under the (obviously sensible) assumption that no object—not even across multiple layers—can contain an object of its own type as a part, this recursive definition is well founded, i.e., it terminates.

This results in a stringent system architecture in the form of a hierarchically layered structure of abstract data types ordered according to increasing complexity of the objects.

The *naming* of the components depends on the programming language used; in Haskell or Modula-2, for example, they are called *module*, in C#, D and Java *classes* and in Go *packages*.

1.1.3 User Manual

Once the system architecture is complete, work begins on an exact description of the system's (external) behaviour and its operation, i.e., the design of the *user interface* and possibly the *interfaces* to peripheral devices.

The user interface is given by the interactions between users (possibly also used peripheral devices) and the automated system running on computers. Its design includes the description of the user inputs into the computers (possibly also through the devices used for data collection) and the outputs of the computers (possibly also the devices intended for data output).

The considerations for this can be divided into two categories:

- the *representation forms* of the objects on the screen and possibly other devices and
- the *inputs* and *commands* for operating the system.

The points mentioned are largely independent of each other: For example, the representation of data on the screen is not dependent on the type of system operation (by keyboard and/or mouse or inputs from data from other devices) and the operation of the program has nothing to do with the display of data on the screen or the outputs on peripheral devices.

As a guide to designing user interfaces, some questions from Nievergelt and Ventura are quoted in [4] that "well characterize most difficulties of users of interactive programs":

- Where am I?
- What can I do here?
- How did I get here?
- Where can I go and how do I get there?

This list of questions must be expanded, e.g., by

- Wie komme ich hier wieder heraus?
- Wie erfahre ich nach einer Unterbrechung der Arbeit, wo ich bin?

- Was *soll* ich hier tun?
- Mit welchen Tasten, Mausklicks, Befehlen o. ä. erreiche ich, was ich will?
- Kann ich etwas *ungeschehen* machen? Wenn ja, *wie*?
- *Welchen* Fehler habe ich gerade gemacht?

The end product of the work in this phase is the user manual (often also called "operating instructions").

1.1.4 Construction

The construction of a program (see Fig. 1.5) in the sense of the program life cycle consists according to the predicted of two parts:

- for each component from
  - its specification and
  - its implementation
- and the *system integration*—the structuring of the components.

1.1.4.1 Specification of the Components

When a component provides an abstract *data type*, its name (which naturally does not apply to abstract *data objects*) is specified in the "header" of its specification. The name of the data type is either the name of the component, if the syntax of the programming language used allows it, or at least it can be considered a *synonym for it*. This may include a description of the semantics of the data type.

The "body" of a component's specification consists of a list of all *access operations* on the objects (the "variables" of the data type) with their syntax and their semantics, i.e., the specification of their usage prerequisites and effect descriptions. It should not be forgotten to provide operations that allow clients to check the prerequisites.

In the event that there are multiple implementations, which differ, for example, in their efficiency for different usage requirements, of course, correspondingly many different type

Fig. 1.5 Construction

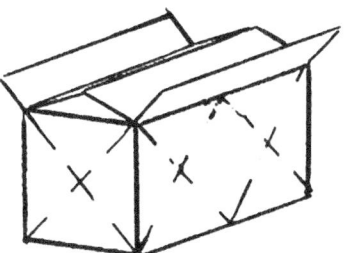

names must be used and constructors specified. In this case, clients should also be appropriately informed by commenting on the types.

If necessary, *constants* are specified, for example, as limits for certain ranges, or for naming the components of enumeration types such as in a data type *calendar date weekday* (monday, tuesday, …) and *period* (daily, weekly, monthly, …).

Variables, on the other hand, due to the dangers associated with uncontrollable changes from the outside, should *under no circumstances* be included in the specification.

Operations are provided for the manipulation of *component variables* (i.e., global variables in the implementations of the components that are not accessible from the outside), which deliver or change their values.

1.1.4.2 Implementation of the Components

In the implementation of a component, the first step is to determine the concrete data type that models the specified data type, or the concrete representation of the data object that is the carrier for the operations on the abstract data object. This usually involves one of the following cases:

At the lowest level by

- the choice of an elementary data type (character, string, truth value, natural or integer number, floating point number, self-defined enumeration type);

and at higher levels by

- the construction of new objects, whose attributes are given objects of different types, by "binding" them together by means of a type constructor "tuple"; or
- the grouping of given objects into sets of objects (which are themselves objects), for example, in static or dynamic fields, scatter storage tables, in dynamic meshes (lists, trees, graphs, or in persistent type constructions for permanent storage (sequential or indexed sequential files, B-trees or similar).

The implementation of the operations then often involves well-known algorithms for processing the respective data structures.

If design errors become apparent during the implementation—usually in the form of incompleteness or lack of clarity, possibly due to previously undiscovered contradictions in previous phases—the specification must be corrected in close cooperation with all involved clients and their implementations adapted to the changes.

In our simple model of the program life cycle, no separate phase is provided for *testing the components* no separate phase because we consider tests of the implementations against their specification as part of the implementation.

Fig. 1.6 What *actually* was meant …

1.1.4.3 Component Hierarchy

In simple cases, a program is controlled with an input loop (event loop), in more complex cases with a *selection menu*, through which user inputs branch into individual program parts, which in turn can consist of input loops or selection menus. The *system integration* thus consists of the construction of the input loop(s) and selection menus, incorporating the developed components.

The first step in a revision is to check whether the constructed system (see Fig. 1.6) matches the ideas of the clients or users.

▶ A system does not do what the clients originally *imagined*, but what the developers *constructed*.

The goal of a system revision is then, if necessary, appropriate corrections and usually an extension of the system's functionality or an adaptation to changed conditions for its use. It therefore consists of a re-entry into the first phase of the program life cycle, from where it is cycled through again.

1.2 Advantages of an Object-Based System Architecture

It is easy to see that the demands on the components of a system's decomposition according to its objects out from Sect. 1.1.2.1 and that all postulates from Sect. 1.1.2.2 for a proper component concept are naturally fulfilled.

1.2.1 On Specification

Unfortunately, a decisive weakness of many common programming languages becomes apparent here: Lack of syntactic support for safeguards against non-compliance with prerequisites or for assurances of effects. An optimal language level is of course an algebraic specification of the operations by equations (relationships between the operations).

1.2.2 On Implementation

The *understandability and clarity of the specification* and the minimality of its service offering are guaranteed a priori by the treatment of exactly one data type (possibly data object).

The minimality of the service offering is ensured because it only concerns access to objects of *one* data type; the same applies to the independence from the specification of other components, as far as the specification does not extend that of other data types. The reduction of data transports is trivially achieved because only operations on the abstract data type with their parameters are provided.

When using *abstract data types*, the invisibility of the structure of an object from components in the specification automatically ensures the preservation of the *principle of secrecy* and a maximum independence from other components.

Degree of generality and *completeness* result in connection with a desired maximality of the service offering within the set limits: in any case, a sufficiently large variety of accesses to the objects of the considered data type should be provided in order to make the component as universally usable as possible. This does not contradict the principle of *openness*: A component can always be extended by specifying and implementing initially unconsidered, but later recognized as necessary accesses (which of course requires the recompilation of its clients).

1.2.3 On Implementation

The postulates considered necessary for the *implementation* are, in a sense, fulfilled by the implementation:

The *manageability* of the emerging complexity is ensured by the construction of data types from components that have been previously defined. Since their construction details are hidden in other implementations, the implementation of the composite data types does not have to worry about details, but can assume their existence and the accesses to them only on the basis of the knowledge of their definition, i.e., on a rather abstract level.

If it turns out during the implementation that further, initially unforeseen parts are necessary, this gives rise to the construction of separate components, which are then used—again only based on their abstract description, i.e., their specification.

The *principle of secrecy* can be excellently exploited: can be excellently exploited:

The replacement of implementations by alternatives is optimally supported by the described principle. Typical examples—mind you, with the same specifications—can be found in the following scenarios:

Different implementations may be required depending on whether data stocks are only recorded once and then preferably searched, or whether they are continuously updated and research is comparatively rare; the implementation of access to data in *one computer* fundamentally differs from access to distributed data located on *different computers* (see [1]).

Often it is necessary to examine alternatives that balance between contradictory requirements for favourable runtime behaviour of a component and the demand for minimal memory requirements. Accesses to the base machine are isolated in suitable components, whose implementations can differ significantly from each other for different target systems.

The *interference freedom* is ensured by the independence of the specification and implementation parts of the components.

To avoid the use of system-wide state information, which is visible throughout the system, thus also manipulable—and therefore almost inevitably a source of highly fatal, but hard to find errors in larger systems—*global variables*—which can only be referred to as *devil's stuff* (!) for this very reason—must be encapsulated in components (possibly in local subcomponents). This keeps their content—unlike local variables in operations—preserved throughout the entire program runtime, but they are safe from uncontrollable external access. Of course, this assumes that the programming language used supports such a concept.

Other useful aspects of the described method are that it

- allows for a stringent further development of a prototype by refining the structures that have occurred so far or by combining them with other structures into larger units;
- basically provides some assurance that reusable parts of program systems are constructed (which is very valuable given the costly development work in the software sector); and
- does not stand in the way of largely decentralized program development.

References

1. Dijkstra, E. W.: The Humble Programmer. Commun. ACM 15 (1972), 859–866. https://www.doi.org/10.1145/355604.361591. https://www.cs.utexas.edu/users/EWD/ewd03xx/EWD340.PDF
2. Liskov, B., Guttag, J. V.: Abstraction and Specification in Program Development. The MIT Press – McGraw-Hill Book Company (1986)
3. Maurer, Ch.: Nonsequential and Distributed Programming with Go. Springer-Verlag (1921) ISBN 978-3-658-29781-7
4. Nievergelt, J., Ventura, A.: Die Gestaltung interaktiver Programme – mit Anwendungsbeispielen im Unterricht. B. G. Teubner, Stuttgart (1983)
5. Parnas, D. L.: A Technique for Software Module Specification with Examples. Commun. ACM 15 (1972), 330–336. https://www.doi.org/10.1145/355602.361309
6. Parnas, D. L.: On the Criteria To Be Used in Decomposing Systems Into Modules. Commun. ACM 15 (1972), 1053–1058.
https://www.doi.org/10.1145/361598.361623

Aspects of Go

> *Is Go an object-oriented language?*
> *Yes and No.*
>
> The Go Authors
> http://golang.org/doc/go_faq.html#Is_Go_an_object-oriented_language

Abstract

This chapter does not provide a complete introduction to Go, in particular, it is not intended to replace the language description2E Rather, it deals with aspects of Go that are not directly apparent from the documentation of the Go developers, namely, how fundamental software engineering principles can be implemented in Go. The package concept is explained and it is shown how abstract data types can be realized as packages. This is followed by considerations on the concept of variables or objects and the distinction between value and reference semantics—a central point in OOP.

Work on the design of the Go programming language began at Google in the fall of 2007; Go was released in November 2009; the first stable version Go 1 was released at the end of March 2012.

Go allows programming at various levels of abstraction:

- from the lowest possible
  - by integrating parts of programs written in Assembler or C, and system libraries;
- to the simplest
  - for the development of small programs; and
- up to the highest
  - by grouping entire groups of components into components, through which abstract design patterns are realized.

2.1 About the Installation of Go

Compilers, source codes, license terms, etc. can be found on the World Wide Web (see [1]). Instructions for installation are stored on the World Wide Web (see [2]). Go requires some environment variables, which are conveniently set in a file in the directory /etc/profile.d. To do this, switch to it as root:

```
cd /etc/profile.d
```

and create the file go.sh there with the following content:

```
export GO111MODULE=auto
export GOOS=linux
export GOARCH=amd64
export GOROOT=/usr/local/go
export PATH=$PATH:$GOROOT/bin
if [ $UID = 0 ]; then
  export GOSRC=$GOROOT/src
else
  export GO=$HOME/go
  export GOPATH=$GO
  export GOSRC=$GO/src
  export GOBIN=$GO/bin
  mkdir -p $GOSRC $GO/pkg $GOBIN
  export PATH=$PATH:$GOBIN
fi
```

These definitions are then valid after every restart of the computer.

The file go.sh is stored on the World Wide Web at

https://maurer-berlin.eu/go and can be downloaded from there.

Go is installed in the directory /usr/local/go by importing the Go repository as root, after deleting any existing older version:

```
cd /usr/local
rm -rf go
tar xfzv go...tgz
```

The third line needs to be specified, e.g., like this:

```
tar xfzv go1.20.2.linuxamd64.tar.gz.
```

The Go library packages are then located in the directory

/usr/local/go/src.

2.2 Packages in Go

The concept of components in Go is that of the package .
There are two types of packages:

- *Program packages*, which implement an executable program, and
- *Library packages*, which provide services for *other* packages.

The source code of a package consists of one or more files, the name of which ends with the suffix ".go". All source code files belonging to a package must be in the same directory and start with the same line

    package ...

where for ...the name of the package is inserted. In the sources of *program packages*, the identifier main must be used, i.e., they must always start with the line

    package main

2.2.1 Program Packages

The source codes of a *program package* can in principle be located in any directory; however, it makes sense to place them in a subdirectory of $GOSRC with the name of the program; source codes of *library packages must* on the other hand be located in a directory (below the node $GOSRC) whose name matches the name of the package.

2.2.1.1 Library Packages
Library packages

- can be distributed across multiple files,
- have a simple export mechanism,
- can have an initialization part, and
- can be nested.

The ability to split a package into multiple files allows in particular the *decomposition* of an *abstract data type*, which is provided by a library package, into

- its *specification* and
- its *implementation(s)*.

Such a separation makes no sense in *program packages* because they do not export anything.

▶ Thus, Go meets the conditions mentioned in Sect. 1.1.2.3

The relationships between packages are regulated by the terms "import" and "export". The syntactic rules for this are very simple:

All identifiers from a package abc that start with a *capital letter* are imported with the instruction:

```
import "abc"
```

imported. Multiple packages are enclosed in brackets and separated pairwise by a semicolon or a line feed. On the other hand, identifiers with a *lowercase* initial letter cannot be accessed from outside the package; their visibility is limited to the source texts within the package (see https://godev.org/ref/spec → Exported Identifiers)).

The import instruction must be at the beginning of a package—immediately after the first line package .... Each imported identifier is then used with the name of the package that exports it as a prefix—separated from the identifier by a dot.

2.2.1.2 Specification of Library Packages

The specification of an abstract data type has the syntactic form of an Interface that begins with the type declaration. After that, both

- the names of used interfaces and
- a list of the signatures of the exported methods that operate on it, and
- possibly additional functions

can follow.

The first of these two cases shows that it can essentially be a recursive definition, which represents a very powerful aspect of Go: Specifications can be "*nested*" in such a way that interfaces—simply by means of the import-clause—can be "*inherited*". This mechanism of "*inheritance*" *at the level of specifications* is, in my opinion, much more significant than those at the level of implementations, because it—when used cleverly—saves mountains of source code lines.

This assessment inevitably sounds quite abstract and can only be understood in the context of suitable examples. However, we will provide many detailed examples for this thesis and explain it in the chapter about the microuniverse using some abstract data types.

At this point, it becomes clear that the design of Go goes far beyond the concept of object-based and realizes a central aspect of object orientation.

A simple example for this:

If the package xyz exports the data type named Xyz and the methods X(), Y() uint, and Z (b bool) and inherits from the interface Abc in the package abc its methods A() and B() which are defined in the package abc, it looks like this:

2.2 Packages in Go

```
package xyz
import "abc"

type Xyz interface {
  Abc
  X()
  Y() uint
  Z(b bool)
}
```

Clients of this package can then use both the methods of Xyz and—without importing the package abc—its methods A() and B() as such on objects of type Xyz.

For a specification of an abstract data type in a package, there can certainly also be alternative implementations, which is very helpful for certain purposes. They realize different design decisions, offering alternatives for clients, e.g., in terms of runtime considerations or memory efficiency. We will also present some examples for this.

The consequence of the facts and postulates presented so far is that Go is excellently suited for object-based software development and allows a rigid implementation of the principle of "*information hiding*", as explained in any textbook on software engineering.

The only restriction, that in Go specifications of abstract dataobjects are not syntactically formulable, can be compensated by constructing a datatype instead of a dataobject, of which only a single instance is created and used.

Only for data objects that encapsulate access to a peripheral device, of which only one instance exists (mouse, keyboard, printer, etc.), this is not sensible, but here one can help oneself with a simple "trick", which will also be presented at a suitable place. However, it must be expressly emphasized here: The advantages of Go's package concept more than compensate for this disadvantage.

2.2.1.3 Constructors

Constructors syntactically have no place in a specification because that—in contradiction to the object-oriented approach—would limit the possible variety of implementations.

But with a simple "trick" this can be circumvented:

A constructor function is included in the specification, which in turn calls a—externally inaccessible, because lowercase—function from the implementation and thus hides the details of its construction. This ensures that clients are informed about the syntax and semantics of the constructors, *without* having to look into the source code of the implementation. (The compulsion to do something like this represents a frequently observed, but highly questionable violation of the principle of "information hiding".) If a package contains several implementations, the constructor functions should contain hints—in the form of comments—about the semantic differences between the corresponding implementations, so that a client can select those constructors that suit his application purposes.

2.2.1.4 Abstract Data Objects

A package can also implement an abstract data *object*, which is useful, for example, when accessing hardware—a computer only has *one* keyboard, *one* mouse, or *one* tty console.

However, it is generally possible to construct abstract data objects using abstract data types. To do this, a data type is defined—only in the implementation—and a single instance of it is created.

In this case, the specification no longer specifies an interface type, but consists only of the access functions to the object "behind the scenes", which constitutes the term *abstract* data object; preferably in the way it was done above with the constructors: The access function in the specification calls a function from the implementation (e.g., with the same name, but a leading lowercase letter).

2.2.2 Packages as Interfaces Only

Packages can also play a different role:

The *recursive* aspect of interfaces mentioned in Sect. 2.2.1.2 naturally suggests that the package concept also makes sense *without* specifying an abstract data type or an abstract data object—simply as a pattern ("*pattern*") for use in other interfaces.

▶ A package can also only define an *interface without* specifying a data type.

An example of this will be given in the following Chap. 4, that of the "objects".

2.2.3 Nesting of Packages

The ability to nest packages proves to be extremely advantageous for the system architecture of larger software systems. A standard small-scale example of this is to "package" separable parts of a package's implementation into a "subpackage", i.e., one that is located in the directory tree below the node of the package.

From a software engineering perspective, this is a significant advantage, as it allows special services of lower layers to be made available for the implementation of packages, which are not readily visible—especially not accessible—from the outside and are thus protected from changes to the specification. We will also provide examples of this at appropriate points.

2.2.4 Initialization of Packages

The initialization part of a *program package* is the body of the "main function" `func main()`, which contains the actual main program; the initialization of a library package consists of the body of the function `init()`.

Both functions have—as the "empty" brackets show—not passed any parameters.

The function `init` is neither exported nor explicitly called, but is executed at runtime of a using program before any function from its package is called. Its task is usually to populate internal (non-exported) data with certain initial values (see https://godev.org/ref/spec/#Program_execution).

If a package contains multiple `init` functions, they run in an unspecified order; the order of execution of the initialization parts in a program that directly or indirectly imports multiple packages is defined by the import dependencies.

2.2.5 Variables of Concrete Data Types

To illustrate basic aspects of object-based programming, let's first summarize the principles of the imperative paradigm that relate to the variable and type concept. By *concrete data types* we mean those data types that are recursively composed of atomic data types using *field, compound, reference, channel, function*, and mapping constructors.

By a *concrete variable* we always mean a variable of a *concrete data type*.

In Go, the following concrete data types exist:

- the *atomic* data types
  `bool` for truth values,
  `int8, int16, int32, int` and `int64` for integers with the synonym `rune` for `int32`,
  `uint8, uint16, uint32, uint, uint64` for natural numbers with the synonym `byte` for `uint8` and `uintptr` for those that represent the value of a pointer (i.e., an address),
  `float32` and `float64` for real numbers,
  `complex64` and `complex128` for complex numbers, and
  `string` for character strings;
- for every concrete data type X and every expression n with the value of a natural number, the array `[n]X`;
- for every concrete data type X the (slice) `[]X`;
- for each sequence X, Y, ... of concrete data types, the compound
  `struct {x X; y Y; ...}` with *components* x of type X, y of type Y, ...;
- for every concrete data type X the *reference type* (= *pointer type*) `*X` with the *dereferencing operator* `*` which assigns to a pointer p of type `*X` the variable `*p` of type X that "p points to" (the choice of the symbol "*" for this operator could be considered somewhat unfortunate, as it already has the meaning of the type also bows to the C world at this point);
- for any two (also empty) sequences X, Y, ..., E, F, ...of concrete data types, the function type `func ([*]X, [*]Y, ...) (E, F, ...`(where the brackets around one result type may be omitted);

- for any (also empty) sequence of interface types or method specifications A, B, ... the interface type `interface A; B; ...`;
- for every concrete data type X, for which equality == (and inequality !=) are defined, and every concrete data type Y the *mapping type* `map[X]Y` as well as
- for every concrete data type X the channel type `chan X`.

For precise syntactic details, please refer to the Go specification

(see https://golang.org/ref/spec#Types).

In the following sections, we will provide detailed explanations with comprehensive examples for all non-atomic data types.

With the declaration of a specific variable x of a data type X `var x X` the following is associated, among other things:

- At the time of the program's translation—i.e., by the compiler—memory space for the value of the variable x is provided, whose size (i.e., "type size" of its type X) is determined by type declaration.
- This memory space is "addressed" under the name x of the variable within its scope, i.e., one can imagine the name of the variable as a *reference* (*pointer*) to the start address of the memory space.
- It is exclusively reserved for its value and is therefore no long available for Q other purposes.
- Its start address is reached under &x.

The need for memory space for a specific variable is given by the *type size* of its data type given. The atomic data types have the following type sizes:

- `bool`, `int8` and `uint8` = `byte`: 1 byte,
- `int16` and `uint16`: 2 bytes,
- `int32`, `uint32`, `float32`: 4 bytes,
- `int` and `uint`: 4 or 8 bytes,
- `int64`, `uint64`, `float64`, `complex64`: 8 bytes and
- `complex128`: 16 bytes.
- `string`: a character string s occupies `len(s)` bytes.

The type sizes of some composite data types can be calculated from this; with

- fields as a product of the value of the constant and the type size of the base type,
- compounds as the sum of the type sizes of their components.

In Go, the memory space requirement of a variable x of a specific data type X is provided by the polymorphic function `Sizeof` from the package `unsafe`.

2.2 Packages in Go

For concrete variables or expressions of specific data types, the usual *standard operations* are provided in Go (while the relevant rules of type compatibility must be observed):

- the value assignment "="
  - to copy the value of an expression into a variable (more precisely: the bit pattern representing the value into the memory space reserved for the variable);
- the equality predicate "==" and its negation "!="
  - to check for matching values of two expressions (more precisely: for bitwise matching of the contents of the memory spaces reserved for them); and
- the predicates of order "<", "<=", ">", and ">="
  - for comparing the sizes of the values of expressions.

Also operations of the package `fmt`:

- `Print, Println, Printf`
  for output on the screen,
- and `Read, ReadString, ReadCard, ReadInt,` and `ReadReal`
  for input via keyboard;
- as well as certain routines from special libraries for querying the mouse for event control with it,
- `&` and the function `Sizeof` from the package `unsafe`
  for accessing the representation of the values of variables as byte sequences in memory via the start address and size of the memory space reserved for them;
- `Read` and `Write` from the package file
  for accessing byte sequences in the file system.

2.2.6 References and Parameters

In this section, the *pointer concept* is examined more closely illuminated, the understanding of which is an indispensable prerequisite for everything else, especially for the realization of the basic concept "object" in *object-oriented programming*.

Go has—just like, e.g., C or Java—only *value parameters*, , *not* however *variable-* (*reference-*)parameters, as they are known from Pascal or Modula-2.

We show here with a simple example how the effect for which variable parameters are used in these languages is achieved just as well with *value parameters*.

The operator +=, with which a variable n of type int is incremented by a value a, could be implemented like this:

```
func inc (p *int, k int) {
  *p = *p + k
}
```

(It should be reminded of the previous section: `*int` denotes the type of references on `int`.)

This function is used with a call in which the start address `&n` is passed instead of the variable `n`:

```
inc (&n, a)
```

This works because the *dereferencing operator* `*` is the inverse operator of the *address operator* `&`:

For variable p of type `*X` and x of type X follows from

```
p == &*p
```

i.e., the value of the pointer p is the starting address of the for x reserved memory space ("p points to x"), that

```
x == *p
```

applies, i.e., x is just the dereferencing of the pointer p.

In particular, it applies (substitute `*p` for x)

```
p == &*p
```

i.e., the value of p is just the starting address of the memory space reserved for `*p`; in short: `*p` is exactly the variable that p points to. Of course, the reverse also applies

```
*&x == x
```

i.e., the variable x is just the dereferencing of its starting address.

For this reason, the instruction `*p = *p + k` in inc has the effect that the pointer p accesses the passed address, from which `Sizeof(int)` bytes are interpreted as the value of a variable of type `int` and changed in *such a way* that this value is incremented by that of the passed expression after the function call. But this now provides *exactly then* the desired effect, when the starting address `&n` of the memory space reserved for `var n int` is passed, which is type-safe due to the signature of the first parameter of `inc`.

The example also teaches us that the naive approach, that value parameters protect against a change of the passed variable, *by no means* applies. But this is not a contradiction, because during the call *not the concrete variable*, but rather *a reference to it* is passed (which is of course *not* changed after the call).

2.3 Variables of Abstract Data Types = Objects

By *abstract data types* we understand those data types, whose existence is secured by the specification of their identifier (and of course their access operations) in a specification, but whose implementation does not need to be known to the clients—the users of the services defined in the specification.

In Go, they are defined in the *definition part* of a package in the form:

```
type X interface { ... }
```

2.3 Variables of Abstract Data Types = Objects

—thus only by specifying their *name*—and are therefore also referred to as opaque data types, because their representation remains "opaque" at this point.

Within the curly brackets, a sequence consisting of

- names of *interface types* or
- names of *methods* with their corresponding *signatures*

are to be specified (we will address the special case of this sequence being empty, i.e., the type `interface`, later on).

In the *implementation part* of the package, they are realized in very simple cases as a *reference* to a concrete data type, otherwise usually as a reference to a compound, whose components *in turn* can be abstract data types.

Analogous to the concrete case, we will henceforth understand an *abstract variable* to always be a variable of an abstract data type.

Abstract variables are essentially declared like concrete ones—but there are two very significant differences:

The value of such a variable is a *reference*, i.e., the address in the working memory from which the value of the variable of the referred type is stored.

Its type size is therefore the address width of the processor of the used computer.

This value is thus to be distinguished from its "actual value" —i.e., the variable, to which it refers—and its type size has *nothing* to do with the type size of the *actual value*.

For this reason, the declaration of such a variable—unlike the procedure carried out by the compiler and runtime system during the initialization of concrete variables—must explicitly follow the provision of memory space for the actual value.

This is, however, a task that is *fundamentally unsolvable* for the compiler:

The concept of separate translatability of specification and implementation results in the "view" of the *actual data* type behind the scenes being impossible, hence the *actual type size* is not known—simply because the existence of the implementation cannot be assumed at this point (which is precisely a major purpose of this independent translatability).

Since the *compiler* therefore *cannot* initiate the reservation of the actually required memory space, this task must be taken over by a *superordinate instance*:

A client of an abstract data type—the person who uses it in a source text—must supplement the declaration of each variable of this type by inserting a statement in which the memory space for the *actual value* is created.

This is conveniently done with functions that return a newly created variable of the relevant abstract data type as a value. They can be equipped with parameters for certain purposes.

The functions that accomplish this are called *constructors* in object-oriented programming.

This is a characteristic feature of *object-oriented programming*.

From now on, we will refer to *abstract variables* as *objects*. Conclusion:

▶ Objects must be explicitly created before they can be processed.

A second important point is that in Go the type name in the implementation must not be the same as in the specification (this has, among other things, system-immanent reasons, which we will discuss later).

Thus, the declaration of a variable x of an abstract data type ABC, provided by a package abc, reads

```
var x abc.ABC
```

and the object x is created with the assignment

```
x = abc.New()
```

These two lines can also be combined into one declaration:

```
var x abc.ABC = abc.New ()
```

or—even shorter—by taking advantage of Go's dynamic type adaptation:

```
x := abc.New ()
```

2.4 Value Versus Reference Semantics

After the considerations from the previous section, the question now arises as to what consequences arise when objects "behind the scenes", i.e., in the implementation, are nothing more than references.

2.4.1 Assignments, Creation of Copies

A value assignment

```
x = y
```

results in *concrete variables* that the value of the variable x is overwritten with that of y. The consequence is that after the assignment there are two different concrete variables with identical value, because different memory locations are reserved for the values of the two variables. Consequently, if the value of the variable y is changed afterwards, the variable x is not affected; its value is *not* changed.

For *objects* x and y, i.e., for variables of an *abstract data type*, this is *not* the case:

With this assignment, the reference y to an object, i.e., merely the address from which the "value" of y can be found, is copied into the reference x. This has a completely different consequence:

The pointer x now refers to the same object as the pointer y, i.e., the variable x now refers to the same object as y. If the object y is changed afterwards, the value of the object x is consequently also changed (in the same way). The first case is an example of *value semantics*, the second for *reference semantics*.

2.4.2 Equality Check and Size Comparison

This distinction should also be made in other cases.

The Boolean expression

```
x == y
```

provides for concrete variables x and y a statement about whether the values of the two variables are equal (*value semantics*), for objects, however, only, whether the pointers x and y refer to the same object (*reference semantics*).

Since the latter does not say anything about the equality of the objects to which x and y refer, we do not get any further with objects.

The situation is even more drastic when it comes to size comparison.

For concrete variables, for whose type the relation < is defined, the Boolean expression

```
x < y
```

provides a statement about whether the value of x is smaller than that of y or not.

For *objects* x and y, on the other hand, it could at best provide the (completely uninteresting) statement, whether the value of the reference x is smaller than that of y, i.e., whether the memory space for the object to which x refers is in the working memory before that to which y refers.

But this is not possible in Go:

The operator < is not defined for references; thus, the approach of comparing objects in terms of size with it is completely unsuitable.

2.4.3 Serialization

For a concrete data type X, with &x you have the start address of a variable x of type X, for a reference p of type *X with *p the specific variable of type X, to which p points, under control; the memory space it occupies is a contiguous area in the working memory, the size of which is known.

For a variable x of an abstract data type X, &x is merely the start address of the value of the reference x—the actual value cannot be found there.

Manipulating the "actual" variable *x would mean accessing the representation details of type X in an implementation bypassing the specification of X. This is not allowed according to the postulated principles of *information hiding*—and in a proper implementation simply impossible, which is achieved by starting the identifiers of the components of the representation of X with *lowercase* letters, thus not being exported.

The type size of a pointer has nothing to do with the size of the memory space of the variable it refers to. This has nothing to do with the actual need for memory space. In particular, it is impossible to access the—in complex cases non-contiguous—areas of the working memory where an object is stored.

The consequence is that objects can *neither* be stored as a sequence of bytes in a file *nor* sent as such over the network.

Access to concrete variables is via their names and the memory space for their value is provided by the translator by declaration, because their size is determined by the type specification. They are processed by value semantics.

With *abstract variables*, i.e., *objects*, things are quite different:

They are accessed via *references*.

For the reasons explained in detail, reference semantics (with few exceptions under carefully considered conditions) is an unsuitable means for their processing.

To get to objects with *value semantics*, i.e., to achieve the effects that are given with concrete variables by value semantics, the following operations are needed, among others, which cannot be managed with reference semantics are

- for the creation of new objects by constructors (operations for the elimination of no longer used objects are obsolete, because Go has memory cleanup);
- for checking for agreement between objects as well as for making copies of objects;
- for comparing objects with respect to an ordering relation;
- for "emptying" objects, i.e., for deleting their values, as well as for checking whether they are empty;
- for displaying objects on the screen;
- for their interactive changeability (by keyboard, mouse, or similar); and
- for encoding objects as serial byte sequences and vice versa (possibly with the insertion of redundancy for error detection or correction), in order to be able to store them persistently on data storage devices or send them to processes on other computers.

References

1. https://go.dev
2. https://maurer-berlin.eu/go

The Microuniverse   3

> *Object-oriented design is, in its simplest form, based on a seemingly elementary idea.*
> *Computing systems perform certain actions on certain objects; to obtain flexible and reusable systems, it is better to base the structure of software on the objects than on the actions.*
>
> Bertrand Meyer
> *Object-oriented Software Construction*, Prentice-Hall (1988), xiv

Abstract

The microuniverse rigidly implements all principles of object-oriented programming. It consists of many packages with abstract data types and objects for all possible purposes. This chapter introduces some of them that are used in the second part of the book.

Many of these packages originate from my teaching activity in computer science; they were originally written in Modula-2, later converted to Java, and now ported and further developed to Go. First, the central package `obj` is introduced.

This is followed by some principles for constructing *simple user interface*. A significant part of this chapter consists of the presentation of "collections", sets of objects (e.g., sequences, buffers, sets, files, graphs).

3.1 Installation of the Microuniverse

The source codes of the microuniverse can be found on the net at https://maurer-berlin.eu/mU. It makes sense to check there occasionally to see if there is a new version that has fixed errors or is more powerful than previous versions.

The microuniverse is installed either by root in the directory `$GOSRC` or by "users" in the subdirectory `go/src` of their home directory, by unpacking the file μU.`tgz` using the `tar` command with the options `xfz` (for "unpack", "file", "decompress"). This creates—if not already present—the subdirectory μU, where all source texts are stored.

If the *prerequisites* mentioned in the following section are met, the µU library is created with the command `go install` µU

(If µU was previously installed, the directory `$GOSRC/`µU should be deleted beforehand to remove obsolete files.)

For compiling and binding, the script `gi` contained in the directory `$GOSRC/`µU is also very useful, which must be copied into a directory that is included in the path (e.g., `$HOME/bin`).

3.1.1 Prerequisites

A computer with Linux as the operating system, on which Go is installed, is required (see notes on the installation of Go in Sect. 2.1).

All the following information refers to ßopenSUSE indexopenSUSE with the `bash` as the login shell. Other distributions or shells may require adjustments such as "`X=...; export X`" instead of "`export X=...`".

Since essential parts of the microuniverse rely on the following:

- the C library,
- X-Window,
- a font with characters of constant width,
- OpenGL,
- the conversion of graphics between the `ppm` format ("portable pixmap") and other formats (e.g., `gif` or `jpg`),
- prints from within programs.

The following must be installed under ßopenSUSE:

- the basic development environment (tools for compiling and binding applications), especially the GNU C compiler;
- the library for extensions of the X11 protocol `libXext-devel`;
- the `terminus-bitmap-fonts` (http://terminus-font.sourceforge.net);
- the graphics library `Mesa-devel`, which implements the OpenGL specification, (https://www.mesa3d.org);
- the *OpenGL Utility Toolkit*, both in `freeglut-devel` (http://freeglut.sourceforge.net);
- the tool for manipulating graphic formats `netpbm` (http://netpbm.sourceforge.net); and
- the text typesetting system TEX of the genius Donald E. Knuth.

3.1 Installation of the Microuniverse

In other distributions, the corresponding packages may have different names—but this can easily be found out by "googling" these terms. Under ßUbuntu the installation

- of the terminus bitmap fonts and
- the packages necessary for OpenGL

is more complicated; details can be found on my page about the installation of the microuniverse (`https://maurer-berlin.eu/mU/instmU.shtml`).

If these packages are not installed, corresponding error messages will occur when compiling μU.go. After error-free translation, the flawless functioning of all libraries of the microuniverse can be checked with the start of the program μU.

Since the microuniverse makes features available that go far beyond the usual standards of tty-console operation, namely,

- *high-resolution graphic outputs in any colour* and
- the *use of a mouse*

the execution of even demanding event-driven graphic programs *in consoles* is possible. For this, the file /dev/input/mice must be readable for the "world", i.e., have the rights rw-r-r-.

In other distributions, "/input/mice" may need to be replaced by the name of another file—also in the file μU/mouse/def.go.

Access to the "framebuffer", which is necessary for console operation, requires that the file /dev/fb0 has the rights rw-rw-rw-, i.e., read and write rights for all.

If this is too risky for security reasons, you can also ensure that root adds the users to the video group. Both can be secured, for example, by corresponding entries in /etc./init.d/boot.local.

Programs with *graphic outputs* or *use of the mouse cannot* naturally be executed in consoles *via login on remote computers*, but only *on the local computer*, where the mouse and screen are connected, as this accesses these *local* resources.

Under X-Window, i.e., on graphical interfaces, such as KDE, Xfce, Gnome, or IceWM, all programs can be executed *also on a remote computer* if its outputs are *redirected* to the local computer. The easiest (and safest) way to do this is with the secure shell in a window by logging in with the command ssh -X host (host = name of the remote computer) to then start the program there. The prerequisite for this is that the ssh services are installed on the involved computers, the daemon sshd is activated, and one is allowed to log in via ssh on it.

3.1.1.1 Other Operating Systems

Go can also be installed under *Windows*®. But in my attempts to install the microuniverse in this operating system using the Windows Subsystem for Linux (see `https://docs.microsoft.com/de-de/windows/wsl/about`), I encountered a *lot of "traps"*. For this reason, I recommend the use of a virtual machine, e.g., *VirtualBox* (see `https://www.virtualbox.org`), to actually be able to use a Linux distribution.

If I ever manage to adapt to the operating system *MacOS*®—a Unix-system—this will be published on my page on the World Wide Web.

3.1.2 License Terms

The microuniverse is designed solely for use in teaching and therefore has a purely academic character. It provides, among other things, numerous examples and animations for my textbook "Nonsequential and Distributed Programming with Go" (Springer Nature 2021). The sources of the microuniverse can be used without restriction for teaching purposes at universities and schools; however, any form of further use is strictly prohibited.

This software is provided by the authors "as is" and any express or implied warranties, including, but not limited to, the implied warranties of merchantability and fitness for a particular purpose are disclaimed. In no event shall the authors be liable for any direct, indirect, incidental, special, exemplary, or consequential damages (including, but not limited to, procurement of substitute goods of services; loss of use, data, or profits; or business interruption) however caused, and any theory of liability, whether in contract, strict liability, or tort (including negligence or otherwise) arising in any way out of the use of this software, even if advised of the possibility of such damage.

The source codes of the microuniverse are developed with great care and are continuously revised. But there is no error-free software—this naturally also applies to these sources. Their use in programs could lead to damage, e.g., to the burning of computers, the derailment of trains, the meltdown in nuclear power plants, or the crash of the moon. Therefore, the use of any sources from μU in programs for serious purposes is explicitly warned! (Excluded are demo programs for use in teaching.)

Reports of discovered errors and hints on ambiguities are gratefully accepted.

3.1.3 Naming in the Microuniverse

The Microcosm generally uses the following uniform naming:

The naming of the data type or object and the specification of the operations on it, which the corresponding library package exports, is in a file named `def.go`.

The representation of the data and the implementation of the operations bear names that are largely self-explanatory. If there is only *one* implementation (or among several "the" standard implementation), the corresponding file is named after the name of the abstract data type that the package exports.

For example, the package `lockp`, which implements lock synchronization with lock algorithms, contains implementations with the (suggestive) names `dekker.go` and `peterson.go` (Dekker and Peterson are the authors of the two algorithms).

As a consequence of the naming in the microuniverse, we generally use the identifier of the data type in lowercase as the name for the implementation of an abstract data type and—following the conventions of other object-oriented languages—for the constructors usually the identifier `New`. If there are several implementations, further characters are added to these identifiers.

The microuniverse offers a variety of abstract data objects and types in the form of packages. They are *fundamentally* separated in specification and implementation; the abstract data objects are largely equipped with the above-mentioned operations.

Some of these packages are introduced and explained in detail in the following sections. But first, an important note:

To drastically shorten the text in specifications, the following language regulation applies throughout the book:

▶ In all specifications in this book, the *calling* object is *always* referred to as x.

3.2 The Constructor New

As justified in Sect. 2.3 about objects, every object must be *generated* before its use. *Exactly that* is the purpose of the constructors. Their syntax for an abstract data type Abc, which is implemented in a package abc with the filename abc.go, in the microuniverse always basically looks like this

```
func New() Abc
```

In the associated implementation part, the construction of this function—based on type declarations of the kind

```
type abc struct {
            ... ...
                    }
```

always starts with the instruction

```
x := new(abc)
```

which results in memory being allocated for a variable of type abc and its address being entered into the value of *x. The constructor function ends with `return x`. Between these lines, components are possibly set to specific initial values, which are interpreted as empty, as long as they are not to be initialized with the zero values for their type (see Sect. 3.3.3).

The call of functions that manipulate objects or contain them as parameters *always presupposes* that these objects were created by the previous call of the constructor New (or similar). This prerequisite is *indispensable*

The memory space reserved for an object by declaration var x Abc initially has—just like in the concrete case—no defined value, thus contains a *random address*.

Accessing the variable—e.g., with an assignment x = ...—would usually mean accessing an address range in memory that was not "assigned" to the calling process by the operating system. For this reason, the Go runtime system acknowledges every access to an object that was not generated with a panic, i.e., with a program abort and the corresponding error message.

3.3 The Object Package

For many abstract data types in the microuniverse, a number of basic interface types are needed, which will prove to be extremely useful or even necessary.

We gather such things in the central package obj, which we introduce here.

We now proceed step by step and first show its most important "parts", the interfaces

```
Equaler
Comparer
Clearer
Coder
```

Motivation and basis for their construction is the importance that the following interfaces in Java for many classes have–especially partly also for the class Object:

```
java/lang/Cloneable.java
java/lang/Comparable.java
java/util/Collection.java
java/io/Serializable.java
```

We need a basic data type that can hide *all* data types, the *empty interface*. I had previously defined it in the μU/obj package as type: type any = interface{}.

The Go developers apparently understood the purpose of this definition at some point: They included it under the name any as an alias for the empty interface interface in the Go specification.

3.3.1 Equaler

Most objects in computer science can be compared with others to see if they are equal, and they can be copied. As explained in Sect. 2.4 about *value versus reference semantics*, a naive attempt to check two objects x and y for equality with the Boolean expression x == y is as nonsensical as the assignment x = y to copy the object y into the object x, because only references are compared or copied, not the actual objects.

3.3 The Object Package

It should be reiterated that for operations on instances of an abstract data type in the microuniverse, the syntax of method calls is consistently used, as is common in object-oriented programming.

The implementation of the equality check of an object x of type Abc in a package where a data type is constructed is thus encapsulated in a method:

```
func (x *abc) Eq (y any) bool
```

Similarly, the copying of an object in methods

```
func (x *abc) Copy (Y any)
```

or

```
func (x *abc) Clone() any
```

The microuniverse provides both.

This gives us the following interface in the file `equaler.go` with two additional functions:

```
package obj

type Equaler interface {

// Returns true, iff the x has the same type as y
// and coincides with it in all its value[s].
Eq (y any) bool

// Pre: y has the same type as x.
// x.Eq (y) (y is unchanged).
Copy (y any)

// Returns a clone of x, i.e. x.Eq (x.Clone()).
Clone() any

// Returns true, iff a implements Equaler.
func IsEqualer (a any) bool { return isEqualer(a) }

// Pre: a and b are of the same type;
//      both are atomic or implement Equaler.
// Returns true, if a and b are equal.
func Eq (a, b any) bool { return eq(a,b) }

/ Pre: a is atomic or implements Equaler.
// Returns a clone of a.
func Clone (a any) any { return clone(a) }

// Returns true, iff a is atomic or implements Equaler.
func AtomicOrEqualer (a any) bool { return Atomic(a) || isEqualer(a) }
```

For non-mathematicians, it should be noted that the word iff means if and only if.

The function `Atomic` is defined in the file `obj.go` (see Sect. 3.3).

We do not show the implementation of the functions `Eq` and `Clone` here, but refer to the μU-source code for this book.

Even with this first example, clear advantages of the package `object` become apparent:

- Its specifications are valid for all abstract data types that contain the interface `Object`, i.e., those that implement this interface (see remark on the nesting of packages in Sect. 2.2.4 and of specifications in Sect. 2.2.1.3).

- The comprehensibility of larger systems is facilitated by the fact that services with the same semantics in all packages bear the same names—distinguished only by the prefix of the imported package when importing.

3.3.2 Comparer

In computer science, many objects are modelled that are naturally endowed with an ordering relation, such as characters and strings, times, calendar dates, amounts of money, postal codes, sequences, and sets. For types of objects where a predicate for size comparison would make no sense, such as for colours, files, geometric figures, and vectors, this ordering relation is simply the discrete relation, i.e., the equality.

This makes it possible, for example, to sort sequences of them.

A size comparison of objects for which an order is defined, with the operator $x < y$, is, as we have justified, meaningless. Consequently, the check is encapsulated in a method less

```
func (x *abc) less (Y any) bool
```

The corresponding interface in the file comparer.go is

```
package obj

type Comparer interface {

// Pre: x is of the same type as the calling object.
// Returns true, iff the calling object is smaller than x.
Less (x any) bool

// Pre: x is of the same type as the calling object.
// Returns true, iff the calling object is smaller than x
// or equals x.
Leq (x any) bool
}

// Returns true, iff a implements Comparer.
func IsComparer (a any) bool { return isComparer(a) }

// Pre: a and b have the same type;
//      both are atomic or implement Comparer.
// Returns true, iff a is smaller than b.
func Less (a, b any) bool { return less(a,b) }

// Pre: a and b have the same type; both
//      both are atomic or implement Comparer and Equaler.
// Returns true, if a is smaller than b or a equals b.
func Leq (a, b any) bool { return leq(a,b) }
```

3.3.3 Clearer

It is generally useful to enrich the "value set" of objects with an "empty" value, i.e., to allow *empty objects*. They can—of course depending on the context of the semantics of the respective type—be interpreted as "undefined", "unknown", etc. What "empty" means

3.3 The Object Package

depends on the semantics of the type of the calling objects. If it is a set or sequence, the meaning is clear; otherwise, it is, for example, an object with an undefined value, represented by a text consisting only of spaces.

This also subsumes the *input* of new objects under the concept of "*editing*" (changing) of values, by, for example, overwriting the empty strings. A newly created object should definitely be empty.

To check whether an object has an *empty* value is served by a method

```
func (x *abc) Empty() bool
```

and overwriting an object with *empty values* is served by the method

```
func (x *abc) Clear()
```

The corresponding operations are specified in the second interface type of Object in the file clearer.go:

```
package obj

type Clearer interface {

// Returns true, iff the calling object is empty. What "empty"
// actually means, depends on the very semantics of the type
// of the objects considered. If that type is, e.g., a collector,
// empty means "containing no objects"; otherwise it is normally
// an object with undefined value, represented by strings
// consisting only of spaces.
Empty() bool

// The calling object is empty.
Clr()
}

// Returns true, iff a implements Clearer.
func IsClearer (a any) bool { return isClearer(a) }

// a is empty.
func Clear (a any) any { return clear(a) }
```

3.3.4 Coder

Objects can be transformed into (in memory contiguous) unstructured byte sequences, the interpretation of which at the machine level as values of objects is not possible, in order to store them persistently on an external storage, for example, or to transport them as "data packets" over the network to processes on other computers.

For this purpose, an object must be "encodable" (also called "serializable") and "decodable", i.e., it must be able to be uniquely restored from a byte sequence into which it was encoded.

The appropriate type for such byte sequences is *slices* of bytes ([]byte), to which we give our own name ("*Stream*"):

```
package obj

type (
Stream = []byte
anyStream = []any
UintStream = []uint
)
```

Such *"type aliases"* have been included in the specification of Go with Go 1.9.

For this purpose, two methods are provided: one for encoding

```
func (x *abc) Encode () []byte
```

and one for decoding

```
func (x *abc) Decode (b []byte)
```

Two prerequisites must be observed:

In the implementation of Encode, the `slice` must be provided to accommodate the value to avoid memory access errors with a call to the make function. For this purpose, a method for announcing the required number of bytes for encoding

```
func (x *abc) Codelen ( ) uint
```

is needed.

The prerequisite for calling Decode is, of course, that the byte sequence b with len(b) == x.Codelen() represents an encoded object of the same type as the caller. The interface can be found in the file coder.go

```
package obj

const
C0 = uint(8) // == Codelen(int(0)); == Codelen(uint(0))
type
Coder interface {

// Returns the number of bytes, that are needed
// to serialise x uniquely revertibly.
Codelen() uint

// x.Eq (x.Decode (x.Encode())
Encode() Stream

// Pre: b is a result of y.Encode() for some y
//      of the same type as x.
// x.Eq(y); x.Encode() == b, i.e. those slices coincide.
Decode (Stream)
}

// Returns true, iff a implements Coder.
func IsCoder (a any) bool { return isCoder(a) }

// Pre: a is atomic or implements Object.
// Returns the codelength of a.
func Codelen (a any) uint { return codelen(a) }

// Pre: a is atomic or implements Object.
// Returns a as encoded byte sequence.
func Encode (a any) Stream { return encode(a) }

// Pre: a is atomic or streamic or implements Object.
//      b is a encoded byte sequence.
```

3.3 The Object Package

```go
// Returns the object, that was encoded in b.
func Decode (a any, b Stream) any { return decode(a,b) }

// Returns a stream of length 16, that encodes a, b.
func Encode2 (a, b int) Stream { return encode2(a,b) }

// Pre: len(s) == 16; s encodes 2 numbers of type int.
// Returns these 2 numbers.
func Decode2 (s Stream) (int, int) { return decode2(s) }

// Returns true, iff a is atomic or implements Coder.
func AtomicOrCoder (a any) bool { return Atomic(a) || isCoder(a) }

// Pre: For each i < len(a): c[i] == Codelen(a[i]).
// Returns a encoded as Stream.
func Encodes (a anyStream, c []uint) Stream { return encodes (a,c) }

// Pre: For each i < len(a): c[i] == Codelen(a[i]).
//      s is an encoded anyStream.
// a is the anystream, that was encoded in s.
func Decodes (s Stream, a anyStream, c []uint) { decodes (s,a,c) }
```

We will not go into the implementation of the functions `Codelen`, `Encode`, and `Decode` here, but refer to the tools from the packages `asn1`, `json`, and `gob` from the Go package `encoding` or our simple constructions in the packages of the microuniverse.

3.3.5 The Interface of the Package Obj

Topics such as *genericity* or *parametric polymorphism* are not covered in this book, because as will be shown, this is much simpler in Go. A crucial contribution to this is the specification of the data type, which will now be introduced at the end of this section.

Strongly influenced by the ideas that revolve around the root of the class hierarchy in Java, the class "`Object`", it makes sense to define an interface in Go that "defines" *objects*.

Every "reasonable" object should—for the reasons stated in the respective interface—implement all four of the above-mentioned interfaces. Excluded from this requirement are of course "atomic objects", i.e., variables of simple data types (see `func Atomic`).

```go
package obj

type Object interface {

// Most objects in computer science can be compared with others,
// whether they are equal, and can be copied, so they have the type
Equaler // see equaler.go

// Furthermore, usually we can order objects; so they have the type
Comparer // see comparer.go

// Moreover they can be empty and may be cleared with the effect
// of being empty, hence they have the type
Clearer // see clearer.go

// and can be serialised into connected byte sequences,
// e.g. to be written to a storage device or transmitted
// over communication channels, so they have the type
Coder // see coder.go
}
```

```
// Returns true, iff the type of a is bool,
// [u]int{8|16|32}, float[32|64], complex[64|128] or string.
func Atomic (a any) bool {
if a == nil {
return false
}
switch a.(type) {
case bool,
int8,   int16,   int32,    int,    int64,
uint8,  uint16, uint32,   uint,   uint64,
float32, float64, complex64, complex128,
string:
return true
}
return false
}

// Returns true, iff the type of a implements Object.
func IsObject (a any) bool {
if a == nil {
return false
}
_, o := a.(Object)
//  _, e := a.(Editor) // Editor implements Object
return o // || e
}

// Returns true, iff a is atomic or implements Object
// (the types that are particularly supported by µU).
func AtomicOrObject (a any) bool {
return Atomic (a) || IsObject (a)
}

// Returns true, iff a is atomic, streamic or implements Object
// (the types that are particularly supported by µU).
func AtomicOrStreamicOrObject (a any) bool {
return Atomic (a) || Streamic(a) || IsObject (a)
}
```

Since it makes sense to package non-atomic variables into abstract data types, this "classification" into

- atomic variables and
- objects

is quite consistent.

All these so far developed interfaces, methods, and functions are used in many other data types that we introduce in this chapter. For reasons of simplification, there are a number of other interfaces in obj. We will introduce some of them in the following, without explaining much about each—the specifications provide enough information about their meaning. Further interfaces from this package will be introduced in the context of the packages for which they are needed.

3.3.6 Stringer

Various objects in computer science can be uniquely identified by strings. For this purpose and to check whether a string represents an object, the following interface is used in the file stringer.go:

```
type Stringer interface {
```

3.3.7 Formatter

This interface in the file formatter.go is needed for data that can be represented differently, like, for example,

- Calendar dates—with or without weekday—25.7.2024, 25.07.24 or Thursday, 25 July 2024—or
- Time specifications—with or without seconds—17.30, 17.30:20.

```
package obj

type Format byte
type Formatter interface {

// Pre: f < Nformats of the objects of the type of x.
// x has the format f.
SetFormat (f Format)

// Returns the format of x.
GetFormat() Format
}

func IsFormatter (a any) bool {
if a == nil { return false }
_, ok := a.(Formatter)
return ok
}
```

3.3.8 Valuator

This interface in the file `valuator.go` is used, for example, for rated graphs whose vertices or edges have values.

```
package obj

type Valuator interface {

// Returns the value of x, if IsValuator (x).
// Returns otherwise 1.
Val() uint

// Pre: IsValuator (x).
```

```
// x.Val() == n (1 << a, if x has the type uint<a>).
SetVal (n uint)
}

// Returns true, iff a implements Valuator or has an uint-type.
func IsValuator (a any) bool { return isValuator(a) }

// Pre: IsValuator (a).
// Returns the value of a.
func Val (a any) uint { return val(a) }

// Pre: IsValuator (a).
// x.Val() == n (1 << a, if a has the type uint<a>).
func SetVal (a *any, n uint) { setVal(a,n) }
```

3.4 Input and Output

This section deals with the "human–computer interface". It consists of two parts:

- the *screen* for outputting data and
- the *keyboard* (including the *mouse*) for input and modification of data.

Input and output are conceptually always separated, which is extensively justified in Chap. 1.

The principle of Unix, to bundle all outputs and inputs into the common concept of *files*, has nothing to do with this. Therefore, we present these two parts separately.

3.4.1 Packages for the Screen

Before the screen package `scr` is introduced, we present six packages that are needed for it: those for emphcolours, *fonts* and *screen-* or *window sizes*, and the three small ones with constants for *cursor shapes*, *line thicknesses*, and appearances of the *mouse pointer*.

3.4.1.1 Colours

The definitions of colours and a number of methods for their manipulation are encapsulated in the data type `Colour` in the package `col`.

```
package col
import . "µU/obj"
type Colour interface {

Object // empty colour is black

// Encodes the colour with red and blue reversed.
EncodeInv() Stream

// String returns the name of x, defined by the name given with New3.
Stringer
```

3.4 Input and Output

```
// String1 returns (rrggbb", where "rr", "gg" and "bb" are the rgb-
    values
// in sedecimal basis (with uppercase letters).
String1() string

// Defined returns true, iff s is a string of 3 values in sedecimal
    basis
// (with uppercase letters). In that case, c is the colour with
// the corresponding rgb-values; otherwise, nothing has happened.
Defined1 (s string) bool

// Return the values of red/green/blue intensity of x.
R() byte; G() byte; B() byte

// x is the colour defined by the values of r, g and b.
Set (r, g, b byte)
SetR (b byte); SetG (b byte); SetB (b byte)

// Liefert x.R() + 256 * x.G() + x.B().
Code() uint

// Returns true, if c is, what the name of the func says.
IsBlack() bool
IsWhite() bool
IsFlashWhite() bool

// Returns the rgb-values of x scaled to the range from 0 to 1.
Float32() (float32, float32, float32)
Float64() (float64, float64, float64)

// c is changed in a manner suggested by the name of the method.
Invert()
Contrast()
}

// Returns the colour Black.
func New() Colour { return new_() }

// Returns the colour defined by (r, g, b).
func New3 (r, g, b byte) Colour { return new3(r,g,b) }

// Returns the colour defined by (r, g, b) with name n.
func New3n (n string, r, g, b byte) Colour { return new3n(n,r,g,b) }

func HeadF() Colour { return flashWhite() }
func HeadB() Colour { return blue() }
func HintF() Colour { return flashWhite() }
func HintB() Colour { return magenta() }
func ErrorF() Colour { return flashYellow() }
func ErrorB() Colour { return red() }
func MenuF() Colour { return flashWhite() }
func MenuB() Colour { return red() }

// Returns a random colour.
func Rand() Colour { return random() }

// Returns the fore- and backgroundcolours at the start of the system
// for unmarked and marked objects.
func StartCols() (Colour, Colour) { return startCols() }
func StartColF() Colour { return startColF() }
func StartColB() Colour { return startColB() }

// Returns (FlashWhite, Black).
func StartColsA() (Colour, Colour) { return startColsA() }

// Returns the slice of all colours defined in this package.
func AllColours() []Colour { return allColours() }
```

```
func Black() Colour      { return black() }
...
many further colours
...
```

The colours (e.g., Red(), Orange(), Yellow(), Green(), Cyan(), Blue(), Magenta()) are defined as methods to protect them from external access.

3.4.1.2 Fonts

There are only a few fonts needed for the *screen* and for the *printing of texts*. The default is 8×16, which corresponds to the standard resolution in console operation. For the fonts, the *terminus-fonts* (see Sect. 3.1.1) are used, which are not proportional, which is necessary for the *rasterization* explained in the section about the screen (see Sect. 3.4.2).

```
package fontsize

type Size byte; const ( // for prt   for screen
Tiny = Size(iota)   // cmtt6      7 *  5 px
Small               // cmtt8     10 *  6 px
Normal              // cmtt10    16 *  8 px
Big                 // cmtt12    24 * 12 px
Large               // cmtt14    28 * 14 px
Huge                // cmtt17    32 * 16 px
NSizes
)

package font
import "µU/fontsize"

type Font byte; const ( // only for prt
Roman = Font(iota)
Bold
Italic
NFonts
)
const
M = 6 // len names
var
Name []string

// Returns a string of len 2, that uniquely defines f and s.
func Code (f Font, s fontsize.Size) string { return code(f,s) }

// Returns the width resp. the height of a font in size s;
// for prt in pt and for scr in px.
func Wd (s fontsize.Size) uint { return wd(s) }
func Ht (s fontsize.Size) uint { return ht(s) }
```

3.4.1.3 Screen/Window Sizes

With regard to common technical standards, *screen modes* for operation are distinguished, defined in the type Mode with constants of the type byte.

3.4 Input and Output

```
package mode

type Mode byte; const (
None = Mode(iota) // lines x colums for 8x16-font
Mini     //  192 x  160    10 x  24
HQVGA    //  240 x  160    10 x  30
QVGA     //  320 x  240    15 x  40
HVGA     //  480 x  320    20 x  60
TXT      //  640 x  400    25 x  80
VGA      //  640 x  480    30 x  80
PAL      //  768 x  576    36 x  96
WVGA     //  800 x  480    30 x 100
SVGA     //  800 x  600    37 x 100
XGA      // 1024 x  768    48 x 128
HD       // 1280 x  720    45 x 160
WXGA     // 1280 x  800    50 x 160
SXVGA    // 1280 x  960    60 x 160
SXGA     // 1280 x 1024    64 x 160
WXGA1    // 1366 x  768    48 x 171
SXGAp    // 1400 x 1050    65 x 175
WXGAp    // 1440 x  900    56 x 180
WXGApp   // 1600 x  900    56 x 200
WSXGA    // 1600 x 1024    64 x 200
UXGA     // 1600 x 1200    75 x 200
WSXGAp   // 1680 x 1050    65 x 210
FHD      // 1920 x 1080    67 x 240
WUXGA    // 1920 x 1200    75 x 240
SUXGA    // 1920 x 1440    90 x 240
QWXGA    // 2048 x 1152    72 x 256
QXGA     // 2048 x 1536    96 x 256
WSUXGA   // 2560 x 1440    90 x 320
WQXGA    // 2560 x 1600   100 x 320
QSXGAp   // 2800 x 2100   131 x 350
QUXGA    // 3200 x 2400   150 x 400
UHD      // 3840 x 2160   135 x 440
HXGA     // 4096 x 3072   192 x 512
WHXGA    // 5120 x 3200   200 x 640
HSXGA    // 5120 x 4096   256 x 640
HUXGA    // 6400 x 4800   300 x 800
FUHD     // 7680 x 4320   270 x 960
NEW
NModes
)

// Returns the pixelwidth of m.
func Wd (m Mode) uint { return x[m] }

// Returns the pixelheight of m.
func Ht (m Mode) uint { return y[m] }

// Returns the pixelwidth and -height of m.
func Res (m Mode) (uint, uint) { return x[m], y[m] }

// Returns the mode with (w, h) pixels for (width, height),
// if such exists; panics otherwise.
func ModeOf (w, h uint) Mode { return modeOf(w,h) }
```

Default setting in the console Konsolenbetrieb is the one under which Linux is started.

3.4.1.4 Cursor Shapes

For this, the following type is available (in a subpackage of scr):

```
package shape

type Shape byte; const (
Off = Shape(iota)
Understroke
Block
NShapes
)

// Returns the coordinates of the upper left corner and the height
// of the rectangle for all combinations (c, s) with c != s.
func Cursor (x, y, h uint, c, s Shape) (uint, uint) { return cursor(x,y
    ,h,c,s) }
```

3.4.2 Screen

The simplest operation for outputting strings is the built-in function `print[ln`; more powerful is the function `Print[ln]` from the Go package `fmt`. They only allow the output of simple programs without screen masks: With them, texts can only be "rolled" line by line across the screen.

Everything that goes beyond this primitive form of output, such as

- the construction of screen masks and
- the targeted output of texts at defined points of a console or a window

is not easily possible with it and the output of graphical objects is not possible. At the Linux level, there are very complicated, but also very powerful library systems that support such things, e.g., `ncurses` for *console operation* and `Qt`, `gtk` or the libraries from X11 and from OpenGL for *graphical interfaces* (see `http://www.gnu.org/software/ncurses`, `https://www.qt.io`, `https://www.gtk.org`, `https://www.x.org/wiki` and `https://www.opengl.org`). The documentation of these libraries is very extensive and requires a long familiarization, which is immediately apparent from a look at the header files in the subdirectories of `/usr/include/qt5`, in the directory `/usr/include/X11` (with its subdirectories), and in `/usr/include/GLI`.

For the construction of more sophisticated programs, a simple concept for the output is necessary, which "hides" this complicated matter behind an *easily understandable* and *easy-to-use* interface with powerful implementations.

The microuniverse provides in its screen package `scr` the abstract data object `Screen` with a wealth of methods and functions for its management and for the output of texts and graphics. It is essential that these functions work in exactly the same way both in console operation and on graphical interfaces based on X11.

A crucial contribution to the demand for a simple interface for the output of objects is the *merging of the output of text and graphics*. The screen is rastered into text lines and columns or pixel columns and lines. The grid serves to position the output of strings and graphical objects, with the position of methods whose names end with `Gr` referring to pixels.

3.4 Input and Output

A technically very complex subpackage of scr is the mouse package scr/mouse; it is only used for the two implementations scr/console.go for the console and scr/xwindow.go for graphical interfaces under X-Window of the screen package and for the keyboard package kbd (see next Sect. 3.4.3) and should not be used further outside; therefore, it is not presented here.

The package imports, in addition to obj, the packages mentioned in Sect. 3.4.1. It provides methods

- for resetting to normal input in console operation;
- for querying screen parameters;
- for querying and setting screen colours for outputs;
- for clearing and buffering rectangular parts of the screen;
- for manipulating the cursor;
- for outputting characters, strings, and natural numbers;
- for manipulating the font size;
- for outputting simple geometric figures (e.g., points, lines, rectangles, polygons, circles, ellipses, Bezier curves), all also inverted and some also filled;
- for querying the mouse and setting mouse parameters;
- for serializing rectangular parts of the screen;
- for displaying graphic files in ppm format;
- for managing a "clipboard"; and
- for moving in three-dimensional structures that were created using OpenGL

and functions

- for querying the screen size;
- for indicating whether a process is running under X-Window (alternatively on a console); and
- for screen lock synchronization in concurrent programs.

Of course, it also provides *constructors* for creating screens.

In order not to have to constantly put the identifier s of a screen created by, for example, s := NewMax() in front of the method calls from scr, scr also provides an abstract data *object* in the file ado.go.

At the beginning of a program run in *console operation*, the keyboard is set to "raw" input K_MEDIUMRAW. When the program ends, it must be reset to "normal" input K_XLATE (see /usr/include/linux/kd.h) so that the computer remains operable. This is done by the following procedure:

The function init() in kbd/keyboard.go calls the function initConsole in kbd/console.go, in which the function for resetting is added to the set of those functions that are called at the end of a program (see file `halt.go` in the package `ker`). This mechanism is called by the method `Fin()` in the screen package.

At the beginning of a program run, the cursor's blinking is also turned off so that it doesn't annoy; with the call of the method `Fin()`, it is made to blink again.

Thus, the first line of a program that uses the microuniverse typically looks like this:

```
scr.NewMax(); defer scr.Fin()
```

The call of the instruction `scr.Fin()` is absolutely necessary, because otherwise after the end of a program

- in operation under X-Window the cursor is gone and
- in console operation the keyboard is "dead" and therefore the computer is no longer operable.

Here follows the specification, which for obvious reasons is very long, but fulfills the above-mentioned requirement for easy understanding and simple usability:

```
package scr

/* Pre: For use in a (tty)-console:
     The framebuffer is usable, i.e. one of the options "vga=..."
     is contained in the line "kernel ..." of /boot/grub/menu.1st
     (posible values can be found at the end of imp.go).
     Users are in the group video (or world has the rights "r" and "w"
     in /dev/fb0) and world has the right "r" in /dev/input/mice.
For use in a screen on a graphical user interface:
     X is installed.
Programs for execution on far hosts are only called under X.
Fore-/background colour of the screen and actual fore-/backgroundcolour
are FlashWhite and Black. The screen is cleared and the cursor is off.
In a console SIGUSR1 and SIGUSR2 are used internally and not any more
     available.
No process is in the exclusive possession of the screen. */

// #cgo LDFLAGS: -lX11
// #include <X11/Xlib.h>
import
"C"
import (
"µU/env"
"µU/obj"
"µU/col"
"µU/mode"
. "µU/fontsize"
"µU/font"
"µU/scr/shape"
"µU/linewd"
)
type
Event struct {
          T,    // type
          C,    // xkey.keycode, xbutton.button, xmotion.is_hint
          S uint // state
}
var
Eventpipe chan Event = make (chan Event) // only for XWindow
const ( // mousepointer representations (see /usr/include/X11/
     cursorfont.h)
```

3.4 Input and Output

```
Crosshair =  34
Gumby     =  56
Standard  = 132 // top_left_arrow
)
```
type
Screen **interface** {

```
// The keyboard is switched back to normal mode.
Fin()

// Under X, the screen is newly written.
Flush()

// Under X, in the title bar of the window framing the screen
// the string n appears, unless the screen was initialized by a call of
     NewMax.
Name (n string)

// Returns the actual mode.
ActMode() mode.Mode

// Returns the coordinates of the top left corner of the screen.
X() uint
Y() uint

// Returns - depending on the actual fontsize -
// the number of textlines and -columns of the actual mode.
NLines() uint
NColumns() uint

// Returns the pixelwidth/-height of the screen in the actual mode.
Wd() uint
Ht() uint

// Returns the pixel distance between two textlines
// = charheight/-width of the actual fontsize (s. below).
Wd1() uint
Ht1() uint

// Return the quotient Pixelwidth / Pixelheight of the actual mode.
Proportion() float64

// colours
   /////////////////////////////////////////////////////////////

// The colours of the screen are set to f and b (fore-/background);
// to get the effect of these calls, you have to call "Cls()"
     afterwards.
ScrColours (f, b col.Colour)
ScrColourF (f col.Colour)
ScrColourB (b col.Colour)

// Returns the fore-/backgroundcolour of the screen.
ScrCols() (col.Colour, col.Colour)
ScrColF() col.Colour
ScrColB() col.Colour

// The actual foregroundcolour is f, the actual backgroundcolour is b
// resp. that of the screen.
// The colours of the screen are not changed.
Colours (f, b col.Colour)
ColourF (f col.Colour)
ColourB (b col.Colour)

// Returns the actual fore-/backgroundcolour.
Cols() (col.Colour, col.Colour)
ColF() col.Colour
```

```
ColB() col.Colour

// Returns the colour of the pixel at (x, y).
Colour (x, y uint) col.Colour

// ranges
   /////////////////////////////////////////////////////////////

// Pre: c + w <= NColumns, l + h <= NLines resp.
//      x <= x1 < Wd, y <= y1 < Ht.

// The screen is cleared between line l and l+h and column c and c+w
// (both including) in its backgroundcolour.
Clr (l, c, w, h uint)

// The pixels in the rectangle defined by (x, y, w, h)
// including) have the backgroundcolour of the screen.
ClrGr (x, y int, w, h uint)

// The screen is cleared in its backgroundcolour.
// The cursor has the position (0, 0) and is off.
// The mouse has the position (?, ?) and is off.
Cls()

// If on, then the screen buffer is cleared and
// all further output is only going to the screen buffer,
// otherwise, the screen contains the content of the screen buffer
// and all further output is going to the screen.
Buf (on bool)

// Returns true, iff the output goes only to the screen buffer.
Buffered() bool

// The content of the rectangle defined by (l/x, c/y, w, h)
// is copied into the archive (the former content of the archive is
//    lost).
Save (l, c, w, h uint)
//   SaveGr (x, y, x1, y1 int)
SaveGr (x, y int, w, h uint)
Save1() // full screen

// The content of the rectangle defined by (l/x, c/y, w, h)
// is restored from the archive.
Restore (l, c, w, h uint)
//   RestoreGr (x, y, x1, y1 int)
RestoreGr (x, y int, w, h uint)
Restore1() // full screen

// cursor
   /////////////////////////////////////////////////////////////

// Pre: l < NLines, c < NColumns.
// The cursor has the position (line, coloumn) == (l, c)
// and the shape s. (0, 0) is the top left top corner.
Warp (l, c uint, s shape.Shape)

// Pre: x <= NColumsGr - Columnwidth, y <= Ht - Lineheight.
// The cursor has the graphics position (column, line) = (x, y)
// and the shape s. (0, 0) is the top left top corner.
WarpGr (x, y uint, s shape.Shape)

// text
   /////////////////////////////////////////////////////////////

// The position (0, 0) is the top left corner of the screen.
// The pixels of the characters have the actual foregroundcolour,
```

3.4 Input and Output

```
// the pixels in the rectangles around them have the actual
    backgroundcolour
// (if transparency is switched on, those pixels are not changed).

// Pre: 32 <= b < 127, l < NLines, c + 1 < NColumns.
// b is written to the screen at position (line, colum) = (l, c).
Write1 (b byte, l, c uint)

// Pre: l < NLines, c + len(s) < NColumns.
// s is written to the screen starting at position (line, column) == (l
    , c).
Write (s string, l, c uint)

// Pre: x + Columnwidth < Wd resp.
//      x + Columnwidth * Länge (s) < Wd,
//      y + Lineheight < Ht.
// b and s resp., is written to the screen within the rectangle
// with the top left corner (x, y) in the actual colours.
Write1Gr (b byte, x, y int)
WriteGr (s string, x, y int)

// Pre: c + number of digits of n < NColumns, l < NLines.
// n is written to the screen starting at position (line, column) == (l
    , c).
WriteNat (n, l, c uint)
WriteNatGr (n uint, x, y int)
WriteInt (n int, l, c uint)
WriteIntGr (n, x, y int)

// Pre: see above.
// As above, but with fore- and backgroundcolour reversed.
Write1InvGr (b byte, x, y int)
WriteInvGr (s string, x, y int)

// Returns true, iff transparency is set.
Transparent () bool

// Transparence is switched on, iff t == true.
// If it is on, the backgroundcolour is that of the screen.
Transparence (t bool)

// font
    /////////////////////////////////////////////////////////////

// Returns the actual font; at the beginning Roman.
ActFont () font.Font

// f is the actual font.
SetFont (f font.Font)

// Returns the actual fontsize; at the beginning Normal.
ActFontsize () Size

// s is the actual fontsize. NColumns and NLines are changed
    accordingly.
SetFontsize (s Size)

// graphics
    /////////////////////////////////////////////////////////////

// Position (0, 0) is the top left corner of the screen.
// All output is done in the actual foregroundcolour;
// For operations with name ...Inv all pixels have the complementary
// colour of the fgcolour; for operations with name ...Full
// also all pixels in the interior have these colours.
// The actual linewidth at the beginning is Thin.
```

```
// Returns the actual linewidth.
ActLinewidth() linewd.Linewidth

// The actual linewidth is w.
SetLinewidth (w linewd.Linewidth)

// Pre: See above.
// A pixel in the actual foregroundcolour is set at position (x, y)
// on the screen resp. the colour of that pixel is inverted.
Point (x, y int)
PointInv (x, y int)

// Returns true, iff the point at (x, y) has a distance
// of at most d pixels from the point (a, b).
OnPoint (x, y, a, b int, d uint) bool

// Pre: See above.
// At (xs[i], ys[i]) (i < len(xs) == len(ys)) a pixel is set in the
   actual
// foregroundcolour resp. that pixel is inverted in its colour.
Points (xs, ys []int)
PointsInv (xs, ys []int)

// Returns true, iff one of the points at (xs[i], ys[i]) has a distance
// of at most d pixels from the point (a, b).
OnPoints (xs, ys []int, a, b int, d uint) bool

// Pre: See above.
// The part of the line segment between (x, y) and (x1, y1)
// visible on the screen is drawn in the actual foregroundcolour resp.
// the pixels on that part are inverted in their colour.
Line (x, y, x1, y1 int)
LineInv (x, y, x1, y1 int)

// Pre: See above.
// Returns true, iff the point at (x, y) has a distance of
// at most d pixels from the line segment between (x, y) to (x1, y1).
OnLine (x, y, x1, y1, a, b int, d uint) bool

// Pre: See above.
//     If the calling process runs under X:
//          -1<<15 <= x[i], x1[i], y[i], y1[i] < 1<<15
//          for all i < n:= len(x) == len(y).
//     Otherwise:
//          0 <= x[i], x1[i] < Wd and
//          0 <= y[i], y1[i] < Ht for all i < N.
// For all i < n the parts of the line segments between (x[i], y[i])
   and (x1[i], y1[i]),
// that are visible on the screen, are drawn in the actual
   foregroundcolour
// resp. all points on them are inverted.
Lines (x, y, x1, y1 []int)
LinesInv (x, y, x1, y1 []int)

// Pre: See above.
// Returns true, iff the point at (x, y) has a distance of at most d
   pixels
// from each of the line segments between (x[i], y[i]) and (x1[i], y1[i
   ]).
OnLines (x, y, x1, y1 []int, a, b int, d uint) bool

// Pre: See above.
//     x[i] < Wd, y[i] < Ht für alle i < n:= len(x) == len(y).
// From (x[0], y[0]) over (x[1], y[1]), ... until (x[n-1], y[n-1])
// a sequence of line segments is drawn resp. all points on it are
   inverted.
Segments (x, y []int)
```

3.4 Input and Output

```
SegmentsInv (x, y []int)

// Returns true, iff the point at (a, b) has a distance of at most d
   pixels
// from one of the sequence of line segments defined by x and y.
OnSegments (x, y []int, a, b int, d uint) bool

// Pre: See above.
// A line through (x, y) and (x1, y1) is drawn resp. all points on it
   are inverted.
InfLine (x, y, x1, y1 int)
InfLineInv (x, y, x1, y1 int)

// Returns true, iff the point at (a, b) has a distance of
// at most d pixels from the line through (x, y) and (x1, y1).
OnInfLine (x, y, x1, y1, a, b int, d uint) bool

// Pre: See above.
// Between (x, y), (x1, y1) and (x2, y2) a triangle is drawn
// in the actual foregroundcolour resp. all points on it are inverted
// resp. all its interior points (including its borders) are drawn /
   inverted.
Triangle (x, y, x1, y1, x2, y2 int)
TriangleInv (x, y, x1, y1, x2, y2 int)
TriangleFull (x, y, x1, y1, x2, y2 int)
TriangleFullInv (x, y, x1, y1, x2, y2 int)

// Pre: See above.
// Between (x, y) and (x1, y1) a rectangle (with horizontal and
   vertical borders)
// is drawn in the actual foregroundcolour resp. all points on it are
   inverted
// resp. all its interior points (including its borders) are drawn /
   inverted.
Rectangle (x, y, x1, y1 int)
RectangleInv (x, y, x1, y1 int)
RectangleFull (x, y, x1, y1 int)
RectangleFullInv (x, y, x1, y1 int)

// Pre: See above.
// Returns true, iff the point at (a, b) has a distance of at most d
   pixels
// from the border of the rectangle between (x, y) and (x1, y1).
OnRectangle (x, y, x1, y1, a, b int, d uint) bool

// Returns true, iff the point (a, b) is up to distance t
// inside the rectangle given by (x, y) and (x1, y1).
InRectangle (x, y, x1, y1, a, b int, t uint) bool

// The content of the rectangle defined by (x0, y0, x1, y1)
// is copied to the rectangle with the upper left corner (x, y).
CopyRectangle (x0, y0, x1, y1, x, y int)

// Pre: len(x) == len(y).
//      PolygonFull: The polygon defined by x and y is convex (see
   function Convex).
// A polygon is drawn between (x[0], y[0]), (x[1], y[1]), ... (x[n-1],
   y[n-1], (x[0], y[0])
// resp. all pixels on it are inverted resp. the polygon is filled.
Polygon (x, y []int)
PolygonInv (x, y []int)
PolygonFull (x, y []int)
PolygonFullInv (x, y []int)

// Pre: len(x) == len(y).
//      The polygon defined by x and y is not convex (see function
   Convex),
```

```
//       but its lines do not intersect.
//       In a console: (a, b) is a point inside the polygon;
//       under X: The values of a and b do not matter.
PolygonFull1 (x, y []int, a, b int)
PolygonFullInv1 (x, y []int, a, b int)

// Pre: len(x) == len(y).
// Returns true, iff the point at (a, b) has a distance of at most d
    pixels
// from the polyon defined by x and y.
OnPolygon (x, y []int, a, b int, d uint) bool

// Pre: See above. r <= x, x + r < Wd, r <= y, y + r < Ht.
// Around (x, y) a circle with radius r is drawn / inverted
// resp. all points in its interior are set / inverted.
Circle (x, y int, r uint)
CircleInv (x, y int, r uint)
CircleFull (x, y int, r uint)
CircleFullInv (x, y int, r uint)

// Returns true, iff the point at (x, y) has a distance of at most d
    pixels
// from the border of the circle around (a, b) with radius r.
OnCircle (x, y int, r uint, a, b int, d uint) bool

// Returns true, iff the point at (x, y) has a distance of at most d
    pixels
// from the interior of the circle around (a, b) with radius r.
InCircle (x, y int, r uint, a, b int, d uint) bool

// Pre: See above. r <= x, x + r < Wd, r <= y, y + r < Ht,
//      a and b given in degrees.
// Around (x, y) an arc with radius r is drawn / inverted
// resp. all points in its interior are set / inverted
// from angle a to angle a+b, starting at vertical upright position
// with a and b signed in mathematical orientation (counterclockwise).
Arc (x, y int, r uint, a, b float64)
ArcInv (x, y int, r uint, a, b float64)
ArcFull (x, y int, r uint, a, b float64)
ArcFullInv (x, y int, r uint, a, b float64)

// Pre: See above. a <= x, x + a < Wd, b <= y, y + b < Ht.
// Around (x, y) an ellipse with horizontal / vertical semiaxis a / b
// is drawn / inverted resp. all points in its interior are set /
    inverted.
Ellipse (x, y int, a, b uint)
EllipseInv (x, y int, a, b uint)
EllipseFull (x, y int, a, b uint)
EllipseFullInv (x, y int, a, b uint)

// Returns true, iff the point at (A, B) has a distance of at most d
    pixels
// from the border of the ellipse around (x, y) with semiaxis a and b.
OnEllipse (x, y int, a, b uint, A, B int, d uint) bool

InEllipse (x, y int, a, b uint, A, B int, d uint) bool

// Pre: See above. n:= len(xs) == len(ys).
// From (xs[0], ys[0]) to (xs[n], ys[n]) a Beziercurve of order n
// with (xs[1], ys[1]) .. (xs[n-1], ys[n-1]) as nodes is drawn to the
    screen
// resp. all points on that curve are inverted.
// (For n == 0 the curve is the point (xs[0], ys[0]),
// for n == 1 the line between (xs[0], ys[0]) and (xs[1], ys[1]).
Curve (xs, ys []int)
CurveInv (xs, ys []int)
```

3.4   Input and Output

```
// Returns true, iff the point at (x, y) has a distance of at most d
    pixels
// from the curve defined by xs and ys.
OnCurve (xs, ys []int, a, b int, d uint) bool

// mouse
    ////////////////////////////////////////////////////////////////

// The mousepointer is represented by p.
SetPointer (p uint)

// Returns the position of the mouse cursor.
// For the result (l, c) holds 0 <= l < NLines and 0 <= c < NColumns.
MousePos() (uint, uint)

// Returns the position of the mouse cursor.
// For the result (x, y) holds 0 <= x < Wd and 0 <= y < Ht.
MousePosGr() (int, int)

// The mouse position is written to the screen at position (l,c)/(x,y).
WriteMousePos (l, c uint)
WriteMousePosGr (x, y int)

// Pre: The calling process does not run under X.
// The mouse cursor is switched on, iff b (otherwise off).
MousePointer (b bool)

// Pre: The calling process does not run under X.
// Returns true, iff the mouse cursor is switched on.
MousePointerOn() bool

//
//   WriteMousePointer()

// Pre: l < NLines, c < NColumns.
// The mouse cursor has the position (line, column) = (l, c).
WarpMouse (l, c uint)

// Pre: 0 <= x < Wd, 0 <= y < Ht.
// The mouse cursor has the position (row, line) = (x, y).
WarpMouseGr (x, y int)

// Pre: c + w <= NColumns, l + h <= NLines.
// Returns false, if there is no mouse; returns otherwise true,
// iff the the mouse cursor is in the interior of the rectangle
// defined by l, c, w, h.
UnderMouse (l, c, w, h uint) bool

// Pre: 0 <= x <= x1 < Wd, 0 <= y <= y1 < Ht.
// Returns false, if there is no mouse; returns otherwise true,
// iff the mouse cursor is inside the rectangle between (x, y) and (x1,
    y1)
// or has a distance of at most d pixels from its boundary.
UnderMouseGr (x, y, x1, y1 int, d uint) bool

// Pre: 0 <= x < Wd, 0 <= y < Ht.
// Returns false, if there is no mouse; returns otherwise true,
// iff the mouse cursor has a distance of at most d pixels from (x, y).
UnderMouse1 (x, y int, d uint) bool

// serialisation
    /////////////////////////////////////////////////////////

// Pre: 0 < w <= Wd, 0 < h <= Ht.
// Returns the number of bytes, that are needed to serialise the pixels
// of the rectangle between (0, 0) and (w, h) uniquely invertibly.
Codelen (w, h uint) uint
```

```
// Pre: 0 < w, x + w < Wd, 0 < h, y + h < Ht.
// Returns the byte sequence, that serialises the pixels
// in the rectangle between (x, y) and (x + w, y + h).
Encode (x, y int, w, h uint) obj.Stream

// Pre: s is the result of a call of Encode for some rectangle.
// The pixels of that rectangle are drawn to the screen with the upper
      left corner (x, y);
// the rest of the screen is not changed.
Decode (s obj.Stream, x, y int)

// image-operations
   //////////////////////////////////////////////////

WriteImage (c [][]col.Colour, x, y int)

Screenshot (x, y int, w, h uint) obj.Stream

// openGL
   ////////////////////////////////////////////////////////
Go (draw func(), ex, ey, ez, fx, fy, fz, nx, ny, nz float64)
}

func UnderC() bool {
return env.UnderC()
}

func UnderX() bool {
return env.UnderX()
}

// Returns a new screen with the size of the physical screen.
// The keyboard is switched to raw mode.
func New (x, y uint, m mode.Mode) Screen {
if env.UnderX() {
return NewW (x, y, m)
}
return NewC (x, y, m)
}

// Returns a new screen of the size given by the mode m.
// The keyboard is switched to raw mode.
func NewMax() Screen {
if env.UnderX() {
return NewMaxW()
}
return NewMaxC()
}

// Pre: The size of the screen given by x, y, w, h
//      fits into the available physical screen.
// Returns a new screen with upper left corner (x, y),
// width w and height h. The keyboard is switched to raw mode.
func NewWH (x, y, w, h uint) Screen {
if env.UnderX() {
return NewWHW (x, y, w, h)
}
return NewWHC (x, y, w, h)
}

// Returns the (X, Y)-resolution of the screen in pixels.
func MaxRes() (uint, uint) {
if env.UnderX() {
return MaxResW()
}
```

3.4 Input and Output

```
return MaxResC()
}
// Returns true, iff mode.Res(m) <= MaxRes().
func Ok (m mode.Mode) bool {
if env.UnderX() {
return OkW (m)
}
return OkC (m)
}
// Lock / Unlock guarantee the mutual exclusion when writing on the
     screen
// (e.g. to avoid, that a process after having set its colours
// is interrupted in a subsequent draw and later resumes its drawing
// in another colour, that was meanwhile changed by another process).
func Lock() { lock() }
func Unlock() { unlock() }

func Lock1() { lock1() }
func Unlock1() { unlock1() }

func Act() Screen {
if env.UnderX() {
return actualW
}
return actualC
}

// Returns true, iff len(x) == len(y) and x, y define a convex polygon.
func Convex (x, y []int) bool {
return convex (x,y)
}
```

The implementations of `console.go` and `xwindow.go` are very technical, but algorithmically largely uninteresting. The only exception are the algorithms of Bresenham (see [1]), which are used in `console.go` for the output of lines, circles, and ellipses.

3.4.3 Keyboard

The microuniverse encapsulates access to the keyboard and mouse in an *abstract data object*, the package `kbd`, whose implementation uses the package `mouse`.

In addition to the names for the keys, it provides functions

- for reading the keyboard buffer,
- to query whether a mouse exists, and
- to wait for a key press.

For the *character keys*, the alphanumeric keyboard of the computer is available; characters for which no single key is provided, the commands mentioned in the specification are provided.

To operate and control a system with a keyboard and mouse, three groups of *keys* must be distinguished:

- the *alphanumeric keys* for entering strings and numbers;
- the *command keys* for triggering certain system reactions (enter key, backspace key, arrow keys, etc.); and
- the *mouse buttons* and *mouse movements* for navigating on the screen and "clicking" on objects.

Commands triggered with the command keys can be enhanced in their "*depth of effect*" by combining them with suitable prefix keys (the *shift*, *control*, and *meta* keys); each command has a natural number as depth (0 as the base version, increasing numbers for greater depths). This makes it possible in principle to have commands of different depths with the same effect in systems, such as moving in a text to the next character, word, sentence, section, or chapter or in a calendar to the next day, the next week, the next month, year, or decade.

The keyboard and mouse send their characters and commands using the channel concept in Go as *messages* to the screen, which receives and processes them (see variable eventpipe in the specification of scr).

Here is the specification of our keyboard package:

```
package kbd

// >>> Pre: The preconditions of mouse are met.

/* We distinguish between three groups of keys to operate and control a
     system
with keyboard and mouse:
- character-keys (with echo in form of an alphanumerical character
on the screen) to enter texts and numbers,
- command-keys
to induce particular reactions of the system and
- mouse-buttons and -movements
to navigate on the screen.
In order to abstract from concrete keyboards or mouses,
the following commands are provided for the last two groups: */

type
Comm byte; const (
None = Comm(iota)        // to distinguish between character- and
    command-keys,
                         // see specification of "Read"
Esc                      // to leave the system (or a part of it) or to
    reject
Enter                    // to confirm or to leave an input
Back                     // to move backwards in the system
Left; Right; Up; Down    // to move the cursor on the screen and
PgLeft; PgRight; PgUp; PgDown // to move in the system, e.g. in a
    screen mask,
Pos1; End                // in the corresponding direction
Tab                      // for special purposes
Del; Ins                 // to remove or insert objects
Help; Search             // to induce context dependent reactions of the
    system
Act; Cfg;                // and for special purposes
Mark; Unmark             // to mark and unmark objects
Cut; Copy; Paste         // cut buffer operations
Red; Green; Blue         // to handle colours
Print; Roll; Pause       // for special purposes
OnOff; Lower; Louder     // loudspeaker
Go                       // to move the mouse
```

3.4 Input and Output

```
   Here; This; That           // to click on objects and
   Drag; Drop; Move           // to move them around with a mouse
   To; There; Thither         // and to drag and drop them
   ScrollUp; ScrollDown       // for the mouse wheel
   NComms                     // number of commands
)

/* Commands may be enforced in the "depth" of their "impact":
Every command is associated with a natural number as its depth
(0 as basic version, bigger numbers for greater depths).
So we allow for commands with conceptionally equal effects
but variable ranges of "move depth", as e.g. the movement
in a text to the next character, word, sentence, paragraph or page,
or in a calendar to the next day, week, month, year, decade.

Commands of depth 0 are implemented by keys (without metakeys)
or mouse-actions with system independent semantics:
- Enter:                     input-key "Enter"/"Return"
- Esc:                       stop-/break-key "Esc"
- Back:                      backspace-key "<-"
- Left, Right, Up, Down:     corresponding arrow-keys
- PgUp, PgDown, Pos1, End:   corresponding keys
- Tab:                       Tabkey "|<-  ->|"
- Del, Ins:                  corresponding keys
- Help, Search:              F1-, F2-key
- Act, Cfg:                  F3-, F4-key
- Mark, Unmark:              F5-, F6-key
- Cut, Copy, Paste:          F7-, F8-, F9-key
- Red, Green, Blue:          F10-, F11-, F12-key
- Print, Roll, Pause:        corresponding keys
- OnOff, Lower, Louder:      corresponding keys on laptops
- Go:                        mouse moved with no button pressed
- Here, This, That:          left, right, middle button pressed
- Drag, Drop, Move:          mouse moved with corresponding button
     pressed
- To, There, Thither:        corresponding button released

commands of depth > 0 by combination with metakeys:
- depth 1:                   Shift-key,
- depth 2:                   Strg-key,
- depth 3:                   Alt(Gr)-key */

// The calling process was blocked, until the keyboard buffer was not
     empty.
// Returns a tripel (b, c, d) with the following properties:
// Either c == None and the first object from the keyboard buffer
// is the byte b or b == 0 and the first object of the keyboard buffer
// is the command c of depth d.
// This object is now removed from the keyboard buffer.
// If there is no mouse, then c < Go.
func Read() (byte, Comm, uint) { return read() }

// The calling process was blocked, until there is a byte in the
     keyboard buffer.
// Returns the first byte from the keyboard buffer.
// This byte is deleted from the keyboard buffer.
func Byte() byte { return byte_() }

// The calling process is blocked, until there is a command in the
     keyboard buffer.
// Returns the first command and its depth from the keyboard buffer.
// This command is deleted from the keyboard buffer.
func Command() (Comm, uint) { return command() }

// Returns a string, describing the calling Command.
func (c Comm) String() string { return text[c] }
```

```
// Precondition: A byte or command was read.
// Returns the last read byte, if there is one, otherwise 0.
func LastByte() byte { return lastByte() }

// Precondition: A byte or command was read.
// Returns the last read command, if one was read, otherwise None.
// In the first case, d is the depth of the command, otherwise d = 0.
func LastCommand() (Comm, uint) { return lastCommand() }

// c is stored as last read command.
func DepositCommand (c Comm) { depositCommand(c) }

// b is stored as last read byte.
func DepositByte (b byte) { depositByte(b) }

// The calling process was blocked, until until the keyboard buffer
     contained
// one of the commands Enter (for b = true) resp. Esc or Back (for b =
     false).
// This command is now removed from the keyboard buffer.
// Returns true, iff the depth of the command was == 0.
func Wait (b bool) bool { return wait(b) }

// The calling process was blocked,
// until until the keyboard buffer contained command c with depth d.
// This command is now removed from the keyboard buffer.
func WaitFor (c Comm, d uint) { waitFor(c,d) }

// The calling process was blocked, until until the keyboard buffer
// contained one of the commands Enter, Esc or Back.
// This command is now removed from the keyboard buffer.
func Quit() { quit() }

// Returns true, if the keyboard buffer contained one of the commands
// Enter or Here, and false, if it contained one of the commands
// Back or There, for b = false of any depth and for b = true of a
     depth > 0.
// The calling process was blocked, until the keyboard buffer contained
// one of these commands; this command is now deleted from it.
func Confirmed (b bool) bool { return confirmed(b) }
```

3.4.4 Editor

The data type `Editor`, whose specification is housed in the package `obj`, is used for the output and input of objects at defined screen positions:

```
package obj

type Editor interface { // Objects, that can be written to a particular
                        // position of a screen and that can be
                             changed
                        // by interaction with a user (e.g. by
                             pressing
                        // keys on a keyboard or a mouse).
                        //
                        // A position on a screen is given by
                             line- or
                        // pixeloriented coordinates, i.e., by
                             pairs of
                        // unsigned integers (l, c) or integers
                             (x, y),
```

3.4 Input and Output

```
                        // where l = line and c = column on the
                             screen,
                        // x = pixel in horizontal and y =
                             pixel in
                        // vertical direction. In both cases
                             (0, 0)
                        // denotes the top left corner of the
                             screen.
Object

// Pre: l, c have to be "small enough", i.e.
//      l + height of x < scr.NLines, c + width of x < scr.NColums.
// x is written to the screen with
// its left top corner at line/column = l/c.
Write (l, c uint)

// Pre: see Write.
// x has the value, that was edited at line/column l/c.
// Hint: A "new" object is "read" by editing an empty one.
Edit (l, c uint)
}

func IsEditor (a any) bool {
if a == nil { return false }
_, ok := a.(Editor)
return ok
}

type
EditorGr interface {

Editor

// Pre: see above. x, y are pixel coordinates.
WriteGr (x, y int)
EditGr (x, y int)
}

func IsEditorGr (a any) bool {
if a == nil { return false }
_, ok := a.(EditorGr)
return ok
}
```

3.4.5 Input/Output Fields

The output of strings with the function print[ln] (or the more powerful function Print[ln] from the Go package fmt) and their input with the functions Read... from the Go package bufio only allow interaction in programs where screen design does not matter, because they only allow strings to be output and input line by line.

For output and input of strings, fields of defined width within a screen line are provided, from which screen masks can be composed. The microuniverse contains the abstract data type Box for this purpose.

The methods for output and editing are given parameters to determine their starting position on the screen: l, c of type uint for line and column or x, y of type int.

A distinction is made between overwrite and insert mode, which can be switched back and forth. Which mode is switched on can be recognized by the shape of the cursor recognizable: a small cursor (underscore) for the insert mode, a large "block" cursor for the overwrite mode.

The strings in the fields can be provided with certain font and background colours.

Before entering a string in a field, it is pre-set with a defined content (which can also consist only of spaces); this means that there is no need to distinguish between the new entry of strings and their modification.

Each input begins with an output of the field content. When editing, the string output in the field can be changed in a way that is based on common principles—using some of the commands mentioned in the keyboard package. After the input is completed by the provided commands, the field content is handed over to the system, which takes over further control.

The completion of an input is done by commands that do not serve to correct the field content: *Enter*, *Esc*, *Up*, *Down*, *PgUp*, *PgDown*, or other commands in connection with meta keys. With the variety of these commands, it is possible to jump specifically through the fields in a screen mask.

Here is the specification of the data type box:

```
package box
import (. "µU/obj"; "µU/col")

type Box interface { // Boxes within one line of the screen
                                   // to write and edit strings.
Stringer col.Colourer

// Pre: n > 0.
// x has the width n.
Wd (n uint)

// The editor mode is changed to that of a pocket calculator.
SetNumerical()

// See scr.Transparence.
Transparence (t bool)

// Pre: 0 < c < width of the calling box.
// At the beginning, the cursor is in position c.
Start (c uint)

// x has the fore- and backgroundcolour of the screen.
ScrColours ()

// x has the fore-/backgroundcolour f/b.
Colours (f, b col.Colour)
ColourF (f col.Colour)
ColourB (b col.Colour)

// x is filled with an empty string.
Clr (l, c uint)

// Pre: l < scr.NLines,
//      c + width of x <= scr.NColumns,
//      c + len(s) <= scr.NColumns.
//      width of X == 0 or len (s) <= width of x.
// If width of X was 0, now width of x == len(s).
// s is written to the screen,
// starting at position (line, column) == (l, c) in the colours of x.
Write (s string, l, c uint)
```

3.4 Input and Output

```
// Pre: y <= scr.Ht - scr.Ht1,
//      x + scr.Wd1 * width of x < scr.NColumns,
//      x + scr.Wd1 * length of s < scr.NColumns.
// Like Write, starting at pixelpos (column, line) == (x, y).
WriteGr (s string, x, y int)

// Pre: 1 < scr.NLines, c + width of the calling box < scr.NColumns,
//      c + len (s) < scr.NColumns
//      width of x == 0 or length of x <= width of x.
// If width of x was 0, now width of x == len(s).
// s is now the string, that was edited starting at position (1, c).
// To correct while typing, the usual keys can be used:
// - Backspace and Del to remove characters,
//   in combination with Shift or Strg to delete all,
// - arrow keys Left/Right and Pos1/End to move inside x,
// - Ins to toggle between insert mode (underline cursor)
//   and overwrite mode (block cursor).
// The cursor starts at the beginning of x.
// If s was empty, the mode starts with insert, otherwise with
//    overwrite.
// The calling process was blocked, until the input was terminated with
// another command (see kbd) or one of the above commands with depth >
//    0.
Edit (s *string, 1, c uint)

// Pre: y <= scr.Ht() - scr.Ht1(),
//      x + scr.Wd1() * width (of the calling box) < scr.NColumns(),
//      x + scr.Wd1() * len(s) < scr.NColumns().
// Like Edit, starting at pixelpos (column, line) == (x, y).
EditGr (s *string, x, y int)
}

// Returns an new box of width 0,
// the colours of the screen and the default editor mode.
func New() Box { return new_() }
```

3.4.6 Error Messages and Hints

The abstract data object errh is used for the output of

- Error messages and
- User hints.

After an incorrect input, a hint to the error appears in the last screen line. The content of the input field remains standing so that the error can be traced.

The cursor is now not visible. If the acknowledgement of the error message is confirmed with the Esc key, the error text disappears and the cursor reappears at the beginning of the relevant field, with the field editor in overwrite mode, so that the entered string can be corrected.

Here is its specification:

`errh.def.go`

3.4.7 Printer

The package for printing strings is also an abstract data object. A prerequisite for its use is the installation of TeX. It provides functions

- for defining the font,
- for the line and column number based on it on a DIN-A4 page, and
- for printing strings, where their start position (line, column) is given as a parameter.

Here is its specification:

```
package prt

// >>> Pre: TeX is installed.
import ("µU/fontsize"; "µU/font")

var PrintCommand = "lp"

func Possible() bool { return possible() }

// The actual font is f.
func SetFont (f font.Font) { setFont(f) }

// Returns the actual font.
func ActualFont() font.Font { return actualFont }

// The actual fontsize is f.
func SetFontsize (s fontsize.Size) { setFontsize(s) }

// Returns the actual fontsize.
func ActualSize() fontsize.Size { return actualSize }

// Returns the number of lines per page.
func NLines() uint { return nL[actualSize] }

// Returns the number of columns per line.
func NColumns() uint { return nC[actualSize] }

// Spec: See TeX.
func Voffset (mm uint) { voffset(mm) }

// Spec: See TeX.
func Footline (s string) { footline(s) }

// Pre: l < maxL; c + 1 < maxC.
// b is n line l, column c in the actual font and fontsizei
// in the printer buffer.
func Print1 (b byte, l, c uint) { print1(b, l, c) }

// Pre: l < maxL, c + len(s) < maxC.
// s is in line l from column c in the actual font and fontsize
// in the printer buffer.
func Print (s string, l, c uint) { print(s, l, c) }

// All lines of the printer buffer are printed;
// the printer buffer is not empty.
func GoPrint() { goPrint() }

// Pre: n is the name of an postscript file in the actual directory.
```

3.4 Input and Output

```
// This file is printed.
func PrintImage (n string) { printImage(n) }
```

3.4.8 Selections

The package sel provides an abstract data object with functions

- for interactive selection from lists in the form of "pulldown menus" and
- for selecting colours or fonts.

Here is its specification:

```
package sel
import ("µU/col"; "µU/fontsize")

type WritingCol func (uint, uint, uint, col.Colour, col.Colour)

// Pre: 1 < n => m > 0; Z < scr.NLines - 2; w >= 1, c + w <= scr.
   NColumns; i < n.
// The calling process was blocked, until user has selected a value i
   <= n with keyboard or mouse.
// Until then a bar menue of height min(n, scr.NY1 - 1 - 1) and width w
// was written to the screen, starting at line 1, column c,
// consisting of at most h texts >>> TODO <<< with fore/background
   colour b/f,
// one of which has inverted colours (at the beginning this is t[i]).
// Either user has chosen one of the texts with arrow keys or mouse
// (then i < n is now the number, that corresponds to the selected text
   )
// or she has cancelled the selection (then now i == n).
// The bar menue now has disappeared from the screen and its place on
   the screen is restored.
func Select (wc WritingCol, n, h, w uint, i *uint, l, c uint, f, b col.
   Colour) {
select_(wc, n, h, w, i, l, c, f, b)
}

// Pre: 1 < n <= len (T) + 1; l < scr.NLines - 2; w >= 1, c + w <= scr.
   NColumns; i < n.
// The calling process was blocked, until user has selected a value i
   <= n with keyboard or mouse.
// Until then a bar menue of height min(N, scr.NY1 - 1 - 1) and width w
// was written to the screen, starting at line 1, column c,
// consisting of h of the texts t[i] with fore-/background colour b/f,
// one of which has inverted colours (at the beginning this is t[i]).
// Either user has chosen one of the texts with arrow keys or mouse
// (then i < n is now the number, that corresponds to the selected text
   )
// or she has cancelled the selection (then now i == n).
// The bar menue now has disappeared from the screen and its place on
   the screen is restored.
func Select1 (t []string, n, w uint, i *uint, l, c uint, f, b col.
   Colour) {
select1 (t, n, w, i, l, c, f, b)
}

// Returns an interactively selected Colour, true;
// returns Black, false), if the selection was cancelled.
```

```
//func Colour (x, y int) (col.Colour, bool) { return colour(x,y) } //
   28 colours
func Colour (l, c, w uint) (col.Colour, bool) { return colour(l,c,w) }
func Colours (l, c, w uint, cols ...col.Colour) (col.Colour, bool) {
   return colours(l,c,w,cols...) }

// Returns an interactively selected font size.
func Fontsize (f, b col.Colour) fontsize.Size { return size(f,b) }
```

3.4.9 Menues

The abstract data type Menue allows the construction of a menue control for programs, which can be nested arbitrarily.

Here is the specification of the package menu:

```
package menue
import . "µU/obj"

type Menue interface { // Multiway trees of menues and statements.
                      // Each leaf contains a statement, that can
                         be executed;
                      // the other nodes are menues, from which a
                         node or leaf
                      // of the level below them can be selected,
                      // Nodes and leaves are identified by strings
                         .

// If there is a level below x, nothing has happened.
// Otherwise, x is a leaf with statement s.
// While executing s, the name of x appears
// in the top line of the screen, iff t == true.
Leaf (s Stmt, t bool)

// If x is a leaf, nothing has happened.
// Otherwise, y is inserted into the level below x.
Ins (y Menue)

// If x is a leaf, the statement of x was executed and now
// the menue, from which x was selected, is again presented.
// Otherwise, a menue is presented, which allows to select
// a node or leaf from the level below x.
Exec ()
}

// Returns an node with name s without a level below.
func New (s string) Menue { return new_(s) }
```

The type Stmt func() stands for parameterless functions, i.e., for instructions (statement).

3.5 Collections of Objects

By *collections* we understand *entities* of objects of variables of a concrete atomic type (e.g., [u]int.., float.., or string) or of objects of the type Object, whose objects can be imagined as "lined up" and through which one can "move" forward and backward.

These two *directions* are described in parameters with the type bool—true for forward and false for backward.

Each collection either has exactly one *actual* object or its actual object is undefined. A newly created collection is empty, so its actual object is undefined.

Collections include methods

- for removing all objects from a collection, for checking whether objects are contained in the collection, and for indicating the number of objects in it;
- for moving the pointer to the current object;
- for inserting and removing objects;
- for reading the current object;
- for checking whether a certain object is contained in the collection;
- for traversing the collection with an operation (if this operation destroys the order on an ordered collection, it must of course be reordered);
- for merging two collections with type-identical objects (if necessary, maintaining the order by "interlocking"); and
- for ordering a collection and checking whether it is ordered.

These methods are provided by the abstract data type introduced in the following section.

3.5.1 Collector

This interface is also part of the package obj:

```
package obj

// Collections of elements of type object or of variables of
// an atomic type (bool, [u]int.., float.., string, ...)
// in a sequential order.
// Every collection has either exactly one actual element
// or its actual element is undefined.
//
// An order relation is a reflexive, transitive and antisymmetric
// relations r, i.e., for all a, b, c in a collection r(a,a),
// r(a,b) and r(b,c) imply r(a,c), r(a,b) and r(b,a) imply Eq(a,b).
// Furthermore, we consider only linear relations,
// i.e., for all a, b in a collection either r(a,b) or r(b,a).
//
// In all specifications x denotes the calling collection.
//
// Constructors return a new empty collection with undefined actual
    object.
```

```go
type Collector interface {

// Empty: Returns true, iff x does not contain any element.
// Clr:   x is empty; its actual element is undefined.
Clearer

// Returns true, iff the actual element of x is undefined.
Offc() bool

// Returns the nunber of elements in x.
Num() uint

// Pre: a has the type of the elements in x.
// If x is not ordered:
//   If the actual element of x was undefined, a copy of a
//   is appended in x (i.e. it is now the last element in x),
//   otherwise x is inserted directly before the actual element.
// Otherwise, i.e., if x is ordered,
//   If an element b with Eq(b,a) was already contained in x,
//   nothing has changed.
//   Otherwise a copy of a is inserted behind the last element b
//   in x with r(b,a); so x is now still ordered w.r.t. r.
// In both cases all other elements and their order in x
// and the actual element in x are not influenced.
Ins (a any)

// If f and if the actual element of x was defined, then
// the actual element is now the element behind the former actual
// element, if that was defined; otherwise it is undefined.
// If !f and if the actual element of x was defined and was not
// the first element in x, then the actual element of x is now
// the element before the former one; if it was undefined,
// then it is now the last element of x.
// In all other cases, nothing has happened.
Step (f bool)

// If f is empty, the actual element is undefined; otherwise for
// f/!f the actual element of x now is the last/first element of x.
Jump (f bool)

// Returns true, iff for f/!f the last/first element of x is its actual
    element.
Eoc (f bool) bool

// Returns a copy of the actual element of x, if that is defined; nil
    otherwise.
Get() any

// Pre: a has the type of the elements in x.
// If x is not ordered:
//   If x was empty or if the actual element of x was undefined, a copy
      of a
//   is appended behind the end of x and is now the actual element of x
      .
//   Otherwise the actual element of x is replaced by a.
// Otherwise, i.e. if x is ordered:
//   If x was empty, a copy of a is now the only element in x.
//   Otherwise, the actual element in x is deleted and a is inserted
      into x
//   where the order of x is preserved.
Put (a any)

// Returns nil, if the actual element of x is undefined.
// Otherwise, the actual element was removed from x,
// and now the actual element is the element after it,
// if the former actual element was not the last element of x.
// In that case the actual element of x now is undefined.
```

3.5 Collections of Objects

```
Del() any

// Returns true, iff a is contained in x. In that case
// the first such element is the actual element of x;
// otherwise, the actual element is the same as before.
Ex (a any) bool

// Pre: x is ordered.
// Returns true, iff x contains objects b with Leq (a, b).
// In this case, the actual element is the smallest such object,
// otherwise the actual element is the same as before.
ExGeq (a any) bool

// op was applied to all elements in x (in their order in x).
// The actual element of x is the same as before.
// If x was ordered, it is up to the client to check
// if x is still ordered and - if not - to sort x.
Trav (op Op)

// Pre: y is a collector of elements of the same type as x
//      (especially contains elements of the same type as a).
// If x == y or if x and y do not have the same type,
// nothing has changed. Otherwise:
// If x is not ordered:
//   x consists of exactly all elements in x before (in their
//   order in x) and behind them all exactly all elements of y
//   before (in their order in y).
//   If the actual element of x was undefined, now the former
//   first element in y is the actual element of x, otherwise
//   the actual element of x is the same as before.
//   y is empty; so its actual element is undefined.
// Otherwise, i.e. if x is ordered w.r.t. to an order relation,
//   Pre: r is either an order (see collector.go) or
//        r is a strict order and x and y are strictly ordered
//        w.r.t. r (i.e. do not contain any two elements a and b
//        with a == b or a.Eq(b) resp.).
//   x consists exactly of all elements in x and y before.
//   If r is strict, then the elements, which are contained
//   in x as well as in y, are contained in x only once,
//   otherwise, i.e. if r is an order, in their multiplicity.
//   x is ordered w.r.t. r and y is empty.
//   The actual elements of x and y are undefined.
Join (y Collector)

// Returns true, iff x is ordered.
Ordered() bool

// x is ordered.
Sort()
}

func IsCollector (a any) bool {
if a == nil { return false }
_, ok := a.(Collector)
return ok
}
```

3.5.2 Seeker

The abstract data type `Seeker` is designed to specifically access an object at a certain position in its order access within collections. The following interface can be found in the file *seeker.go* in the package `obj`:

```
package obj

type Seeker interface {

Collector

// Returns Num(), iff Offc(); returns otherwise
// the position of the actual object of x (starting at 0).
Pos() uint

// The actual object of x is its p-th object, iff p < Num();
// otherwise Offc() == true.
Seek (p uint)
}

func IsSeeker (a any) bool {
if a == nil { return false }
_, ok := a.(Seeker)
return ok
}
```

The type `type Op func (any)` in the method `Trav` stands for functions with one parameter, i.e., for operations on objects.

The constructor `New` of a collection is given an object—either by an expression with the value of an atomic data type or by an object of type `Object`. This determines the type of objects that can be included in the created collection. The microuniverse contains the following abstract data types that implement the interface `Collector`:

- *sequences,*
- *stacks,*
- *queues,*
- *priority queues,*
- *ordered sets* (*AVL trees with positioning*) and
- *persistent sequences*, i.e., *sequential files,*
- *persistent index sets* (*ISAM files*).

These collections will be introduced in the following sections. But first, we will show two data types that extend `Collector`.

3.5.3 Predicator

This interface in the file `predicator.go` in the package `obj` provides methods

3.5 Collections of Objects

- to specify the number of objects,
- to check whether a certain predicate applies to all objects,
- to search for objects,
- to process only those objects when traversing,
- to offset the pointer to the current object only to those objects,
- to transfer the objects to another collection, and
- to remove all objects,

to which a certain *predicate* applies. The type `type Pred func (any)` stands for predicates, i.e., Boolean functions with one parameter.

Here is the specification:

```
package obj

type Predicator interface {

Collector

// Returns the number of those elements in x, for which p returns true.
NumPred (p Pred) uint

// Returns NumPred(p) == Num(), i.e. returns true, iff p returns true
// on all elements in x (particularly if x has no elements).
All (p Pred) bool

// Returns true, iff there is an element in x, for which p returns true
.
// In that case the actual element of x is for b/!b the last/first such
// element, otherwise the actual element of x is the same as before.
ExPred (p Pred, b bool) bool

// Returns true, iff there is an element in x in direction f
// from the actual element of x, for which p returns true.
// In that case the actual element of x is for f/!f the
// next/previous such element, otherwise the actual element of x
// is the same as before.
StepPred (p Pred, f bool) bool

// Pre: y is a collector of elements of the same type as those in x.
// y consists exactly of those elements in x before
// (in their order in x), for which p returns true.
// The actual element of x is undefined; x is unchanged.
Filter (y Collector, p Pred)

// Pre: See Filter.
// y contains exactly those elements in x (in their order in x),
// for which p returns true, and exactly those elements are
// removed from x. The actual elements of x and y are undefined.
Cut (y Collector, p Pred)

// In x all elements, for which p returns true, are removed.
// If the actual element of x was one of them, now it is undefined.
ClrPred (p Pred)
}

func IsPredicator (a any) bool {
if a == nil { return false }
_, ok := a.(Predicator)
return ok
}
```

3.5.4 Sequences

The microuniverse contains in the package `seq` the abstract data type `Sequence`, sequences of objects of atomic data types or of type `Object`.

The number of objects in the sequence can be arbitrarily large—within the limits of available memory resources.

Sequences can be *ordered*, i.e., they contain their objects in the order given by an *ordering*.

As an idea for the pointer to the actual object in the sequence, imagine the cursor when editing a line of text.

Here is the specification of the sequences:

```
package seq
import . "µU/obj"

type Sequence interface {

Equaler
Coder
Seeker // hence Collector, hence Clearer
Predicator

// Pre: x is not ordered.
// The order of the elements in x is reversed.
Reverse()

// Pre: x is not ordered.
// If x contains at most one element, nothing has happened.
// Otherwise, for b == true, the former last element of x is now the
   first,
// for b == false, the former first element is now the last.
// The order of the other elements has not changed.
Rotate (b bool)
}

// Pre: a is atomic or of a type implementing Object.
// If x contains at most one element, nothing has happened.
// Returns otherwise a new empty sequence with pattern object a,
// i.e., for objects of the type of a.
func New (a any) Sequence { return new_(a) }
```

The *implementation* of the `sequences` relies on a representation as *doubly linked list* of *cells*.

The cells are compounds of

- an *object* of the type of those object that are aufgehoben in the sequences
- and the forward and backward Verzeigerung in the form of two *references* to such cells:

```
type cell struct {
any "content of the cell"
next, prev *cell
}
```

As this example shows, it is also possible in Go not to specify names for the components in a compound, but only their type, whereby it is advisable to specify comments on their semantics enclosed in quotation marks.

3.5 Collections of Objects

For a sequence the following information are held in a compound:

- the *number* of its objects, the *number of the position* of the *actual cell*;
- the references to the *anchor cell* and the *actual cell*; and
- the information, whether the sequence is ordered with respect to the ordering on the objects.

```
type sequence struct {
num, pos uint
anchor, actual *cell
ordered bool
}
```

The actual *sequence* consists of the objects that are stored in memory in the order of the *next*-references from the first cell following the *anchor*.

The anchor carries the object passed to it in the constructor as a "pattern object" to check that only objects of its type are inserted into the sequence. It also serves as a *marker "(sentinel)"* for identifying the beginning and end of the list, and from it, the first cell of the sequence is referenced with next and the last cell with prev.

The inclusion of such a marker node leads to a significant simplification of many algorithms, because case distinctions are easy to make.

The whole construction is briefly referred to as a *doubly linked ring list* with *anchor*.

The inclusion of such a marker node leads to a significant simplification of many algorithms, because case distinctions are easy to make.

The whole construction is briefly referred to as a *doubly linked ring list* with *anchor*.

With the redundant components num and pos in the representation, two invariants are associated:

- The value of num must match the number of cells (excluding the anchor) in the representation of the sequence
- and that of pos with the ordinal number of the actual cell in the list (following the next references), from 0 for the first cell following the anchor to num for the anchor.

Their purpose is to make certain operations more efficient; for example, to determine the number of objects in a sequence, the list does not have to be traversed to count the cells, but the value of num is simply provided in direct access.

Of course, this is associated with the problem of ensuring the maintenance of the invariants when developing the algorithms. This is easy in this example: With each insertion or removal of an object into or from the sequence, num is incremented or decremented.

In this implementation, in the constructor

```
func new_(a any) *sequence {
x := new (sequence)
x.anchor = new(cell)
x.anchor.head = Clone (a)
```

Fig. 3.1 N is to be inserted before A

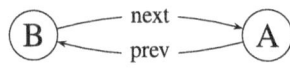

```
x.anchor.next, x.anchor.prev = x.anchor, x.anchor
x.actual = x.anchor
return x
}
```

after providing storage space for the compound sequence and the anchor cell anchor (using new), the anchor is created as the only element of the ring list with a copy of the passed object as content, which points to itself by reference. The anchor is marked as the actual cell (no object is actual) and the invariants num and pos are implicitly set to their zero values, i.e., to 0. Exactly *this* is the representation of an *empty* sequence.

We now demonstrate with two simple examples how typical pointer manipulations are implemented.

An object is *inserted* into an unordered sequence by setting up a cell with a copy of it as content, placing this cell before the actual cell, and incrementing num and pos by 1i. The situation before is the one from Fig. 3.1, where the actual line is marked with A and the following one with B.

An object is *inserted* into an unordered sequence by setting up a cell with a copy of it as content, placing this cell before the actual cell, and incrementing num and pos by 1i. The situation before is the one from Fig. 3.1, where the actual line is marked with A and the following one with B.

The following method is used for this, which causes the newly created cell N to be inserted before the cell that x.actual previously pointed to:

```
func (x *sequence) insert (a any) {
  n := new (cell)
  n.any = Clone (a)
  n.next, n.prev = x.actual, x.actual.prev
  x.actual.prev.next = n
  x.actual.prev = n
}
```

It serves the construction of various other methods, including the method Ins:

```
func (x *sequence) Ins (a any) {
  x.check (a)
  if x.ordered {
    x.actual = x.anchor.next
    x.pos = 0
    for x.actual != x.anchor {
      if Less (x.actual.any, a) {
        x.actual = x.actual.next
        x.pos++
      } else {
        if Less (a, x.actual.any) {
              break
```

3.5 Collections of Objects

Fig. 3.2 N is inserted before vor A

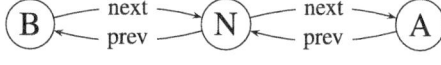

Fig. 3.3 A is to be removed

```
      } else { // Eq (a, x.actual.any), so a is already there
          return
      }
    }
  }
}
x.insert (a)
x.num++
x.pos++
}
```

Afterwards, we have the situation from Fig. 3.2.

The *removal* of an object from a sequence presupposes that the sequence is not empty and that the object to be removed is the actual one. We again denote the actual cell with A and the following one with B and start from the situation in Fig. 3.3.

The actual cell is removed with the following method, which causes the cell that x.actual previously pointed to, is no longer be contained in the sequence and x.actual now points to the cell B, which x.actual.next previously pointed to

```
func (x *sequence) Del() any {
  if x.actual == x.anchor {
    return nil
  }
  c := x.actual.next
  x.actual.prev.next = c
  c.prev = x.actual.prev
  x.actual = c
  x.num--
  return Clone (x.actual.any)
}
```

With this, we have the situation from Fig. 3.4, where B is now the actual cell and cell A will eventually fall victim to Go's garbage collection.

The implementation of various methods has a *linear complexity* in relation to the *number of objects* in the processed sequence, because the list of cells must be traversed. Also, in the sorting method, which is based on the basic idea of *Quicksort*, the sequence must be

Fig. 3.4 A is deleted

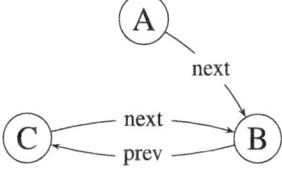

traversed in each recursive step to produce the two parts with smaller or larger objects than the comparison object (the first from the sequence).

At this point, let's leave it at these introductory remarks about the implementation of sequences; the implementation of the other methods will not be discussed here because they are basically trivial.

3.5.5 Stacks

One of the most important structures in computer science is that of a *stack*: a sequence in which objects are stored and from which they are retrieved according to the *"LIFO principle"* (*last in, first out*).

The microuniverse contains in the package stk the abstract data type Stack, unlimited *stacks* of objects of atomic data types or of type Object. Here is its interface:

```
package stk
import . "µU/obj"

type Stack interface { // Not to be used by concurrent processes !

// Returns true, iff there is no element on x.
Empty() bool

// Pre: a is atomic type or of a type implementing object.
// a is the element on top of x, the stack below a is x before.
Push (a any)

// Returns nil, if x is empty, otherwise a copy of
// the element on top of x. That element is removed,
// i.e. x now equals the stack below x before.
Pop() any
}

// Returns a new empty stack for objects of type a.
func New (a any) Stack { return new_(a) }
```

They are represented by sequences:

```
type stack struct { seq.Sequence }
```

and the implementation of the constructor

```
func new_(a any) *stack { return &stack { seq.New(a) } }
```

and the methods are trivial: objects are only inserted and removed at the front (= "top").

3.5.6 Buffers (Queues)

Equally significant are *buffers* (also known as *queues*): sequences into which objects are inserted at the *"FIFO principle"* (*first in, first out*) only at the back and from which they are only removed at the front.

The microuniverse contains in the packages buf and bbuf the two corresponding abstract data types:

3.5 Collections of Objects

- Buffer, *unlimited queues*, and
- BoundedBuffer, *limited buffers* of predetermined (maximum) capacity

of objects of atomic data types or of the type Object.

The specification of the *buffers* is the interface

```
package buf
import . "µU/obj"

type Buffer interface { // FIFO-Queues

// Returns true, if there are no objects in x.
Empty() bool

// Returns the number of objects in x.
Num() uint

// a is inserted as last object into x.
Ins (a any)

// Returns the pattern object of x, if x.Empty().
// Returns otherwise the first object of x
// and that object is removed from x.
Get() any
}

// Pre: a is atomic or of a type implementing Object.
// Returns a new empty queue for objects of the type of a.
// a is the pattern object of this buffer.
func New (a any) Buffer { return new_(a) }
func NewS (a any) Buffer { return newS(a) }
```

There are two implementations of buffers, one with *sequences* and one with *slices*. Both are trivial: When calling Ins(a), a is inserted after the last object of the collection, when calling Get(), the first object is delivered and removed. The method Num does not appear in the implementation with sequences, it is taken over by the used package seq because the data type Sequence implements the type collector—a beautiful example of my thesis of "inheritance at the level of specifications" (see Sect. 2.2.1.2).

Exactly *this* is also shown in the specification of the *limited buffers*, which "inherits" all methods from Buffer:

```
package bbuf
import (. "µU/obj"; "µU/buf")

type BoundedBuffer interface {

buf.Buffer

// Returns true, iff x is filled up to its capacity.
// ! x.Full() is a precondition for a call of x.Ins(a).
Full() bool
}

// Pre: a is atomic or of a type implementing Object.
// Returns an empty buffer of capacity n for objects of the type of a.
func New (a any, n uint) BoundedBuffer { return new_(a,n) }
```

For them too, there are two *implementations*: one in the "classic" form of a *ring buffer* in the form of a circular field, and another one that relies on the buffers. Here is the representation of the first:

```
type boundedBuffer struct {
                                                        any "pattern object"
  cap, num, in, out uint
                                      content anyStream
}
```

Of the second, we show the complete implementation with reference to the above remark:

```
type boundedBuffer1 struct {
                      any "pattern object"
                  cap uint
                      buf.Buffer
                      }
func new1 (a any, n uint) BoundedBuffer {
  x := new(boundedBuffer1)
  x.any = Clone(a)
  x.cap = n
  x.Buffer = buf.New (a)
  return x
}
func (x *boundedBuffer1) Full() bool {
  return x.Num() == x.cap - 1
}
```

3.5.7 Priority Queues

Under *priority queues*, are understood, into which objects of different priority are lined up and from which they are removed in descending order of priority.

The *priority* of the objects is defined by an *order* on them: smaller objects have higher priority. Of course, this assumes that these objects implement Comparer.

The microuniverse contains in the packages pqu and bpqu again two abstract data types:

- *PrioQueue, unlimited priority queues*, and
- *BoundedPrioQueue, limited priority queues* of predetermined (maximum) capacity.

```
package pqu
import (. "µU/obj"; "µU/buf")

type PrioQueue interface {

  buf.Buffer
// Objects are inserted due to their priority, given
// by their order: larger objects have higher priority.
}

// Pre: a is atomic or of a type implementing Object.
func New (a any) PrioQueue { return new_(a) }
```

and the *limited priority queues* through

3.5 Collections of Objects

```
package bpqu
import (. "µU/obj"; "µU/pqu")

type BoundedPrioQueue interface {

pqu.PrioQueue // priority queue with bounded capacity

// Returns true, iff x is filled up to its capacity.
  Full() bool
}

// Pre: a is atomic or of a type implementing Object; m > 0.
// Returns a new empty priority queue for objects of type a
// with maximal capacity m.
  func New (a any, m uint) BoundedPrioQueue { return new_(a,m) }
```

The *priority queues* are represented as a *heaps*, almost perfectly balanced binary trees, whose *lowest leaf layer is always filled from the left* and which fulfill the *heap invariant*:

Every node has the property that the object in it is greater than or equal to the objects in the root nodes of its two subtrees. The root of such a tree contains a largest, i.e., highest priority, object.

The binary trees are implemented in a *slice*, where the positions of the child or parent nodes are found by a simple index calculation:

The root has the index 1, the left or right child node of a parent node with the index i has the index $2i$ or $2i + 1$; the parent node of a node with the index i has the index $i/2$.

The representation of the *priority queues* and the implementation of the constructor look like this:

```
package pqu
import . "µU/obj"

type prioQueue struct {
                   heap []any // heap[0] = pattern object
                   }

func new_(a any) PrioQueue {
  if a == nil { return nil }
  CheckAtomicOrObject (a)
  x := new(prioQueue)
  x.heap = make([]any, 1)
  x.heap[0] = Clone(a)
  return x
}
```

Insertion is done by appending to the last leaf of the lowest layer or—if it is full—insertion as the first object of the layer below with subsequent "rising" by continuous exchange with the node above, as long as the heap invariant is restored.

As an example, we show the insertion of 10 into the heap from Fig. 3.5.

The 10 is placed under the 5 on the right and then 5 and 10 are first swapped and then 10 and 7. This restores the heap invariant and results in the heap in Fig. 3.6.

Here is the implementation of the insertion method:

```
func (x *prioQueue) Ins (a any) {
  CheckTypeEq (a, x.heap[0])
  x.heap = append (x.heap, Clone(a))
  n := uint(len(x.heap))
```

Fig. 3.5 Heap with 12 numbers

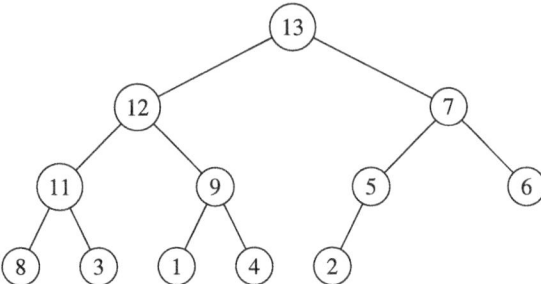

Fig. 3.6 Heap with 13 numbers

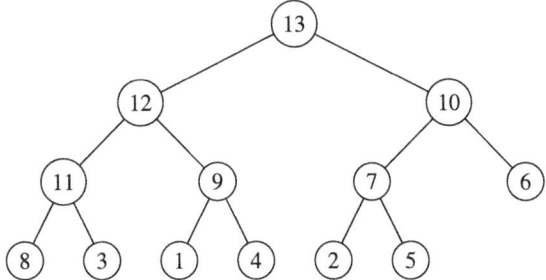

```
    i := n - 1
    for i > 1 && Less (x.heap[i/2], a) {
      x.heap[i] = x.heap[i/2]
      i /= 2
    }
    x.heap[i] = Clone(a)
}
```

When delivering the largest object from the root of the tree, it is removed by replacing the root node with the last node in the lowest leaf layer and this node is deleted from the slice. Then the new root node descends as long as it continues to exchange with the larger of the two nodes below until the heap invariant is restored.

In our example in Fig. 3.6, this results in the heap in Fig. 3.7.

Here is the corresponding implementation:

```
func (x *prioQueue) Get() any {
    if x.Empty() { return x.heap[0] }
    a := x.heap[1]
    if x.Num() == 1 {
      x.heap = x.heap[:1]
      return a
    }
    x.heap[1] = x.heap[x.Num()]
    n := uint(len(x.heap))
    x.heap = x.heap[:n-1]
    x.descend (1)
    return a
}
```

3.5 Collections of Objects

The "descending" of the object from the root to the appropriate place is accomplished with the following recursive method:

```
func (x *prioQueue) Get() any {
  if x.Empty() { return x.heap[0] }
  a := x.heap[1]
  if x.Num() == 1 {
    x.heap = x.heap[:1]
    return a
  }
  x.heap[1] = x.heap[x.Num()]
  n := uint(len(x.heap))
  x.heap = x.heap[:n-1]
  x.descend(1)
  return a
}
```

Both algorithms have the complexity $O(log_2 n)$ for $n =$ number of objects in the queue due to the guaranteed near-perfect balance of the binary trees.

3.5.8 Sets

By *sets* we mean here *fundamentally ordered* sets.

Of course, unlike *sequences*, it is part of the concept that the objects in a set are pairwise different from each other (for elements x and sets M, either $x \in M$ or $x \notin M$ applies, i.e., in particular, for example, $\{x, x\} = \{x\}$). The microuniverse contains for *ordered sets* the package set, which provides the abstract data type Set of atomic variables or objects of the type Object:

```
package set
import . "µU/obj"

type Set interface {

  Equaler
  Collector

// My work is so secret, that even I don't know what I'm doing.
  Write (x0, x1, y, dy uint, f func (any) string)
  Write1 (f func (any) string)
}

// Pre: a is atomic or of a type implementing Object.
// Returns a new empty set for objects of the type of a.
  func New (a any) Set { return new_(a) }
```

Thus, all methods of the type Collector (see Sect. 3.5.1) are available for accessing objects of the type Sequence.

For their implementation, it makes sense to construct algorithms with the most optimal complexity possible when accessing the sets. Of course, *search trees* are suitable for this, i.e., *binary trees*.

However, such trees can also—in the worst case, for example, when many elements are inserted into an empty set in an ordered sequence—lead to completely linear sequences, which counteracts their purpose.

Therefore, it must be ensured that all their nodes always have *as equal as possible* left and right subtrees. A good criterion for "as equal as possible" is the concept of *balance*:

A node in a tree is called

- *balanced*, if it either has no subtrees or the height difference between its left and right subtrees is at most 1;
- *left-heavy*, if its left subtree is higher than its right; and
- *right-heavy*, if its right subtree is higher than its left.

To this end, we define the type Balance:

```
type balance byte; const (
  leftweighty = balance(iota)
  balanced
  rightweighty
)
```

This concept—the AVL trees—was introduced by Adelson and Velskij in their work [2]. They are defined as follows:

- The empty tree is an AVL tree.
- If L and R are AVL trees that differ by at most 1 in height, a tree with a root and with L as the left and R as the right subtree is an AVL tree.

Adelson and Velskij developed algorithms that maintain the AVL property when inserting objects into the trees and when removing objects. Because these algorithms are now the standard for such constructions, they are also used in the implementation of the set package.

The type of nodes in AVL trees is thus clear:

binary trees:

```
type node struct {
              any "content of the node"
     left, right *node
              balance
              }
```

Thus, the representation of the data type looks like this:

```
type set struct {
              any "pattern object"
  anchor, actual *node
              uint "number of objects in the set"
              }
```

The algorithms that maintain the AVL invariant when accessing an AVL tree are

3.5 Collections of Objects

- when *inserting* an object into the tree, *rotations* and
- when *removing* an object from the tree, functions for *balancing*.

We first deal with the insertion and present the necessary *rotation algorithms* using simple examples.

If a 0 is inserted into the AVL tree from Fig. 3.7,

the tree from Fig. 3.8 is created, which violates the AVL invariant.

But with a simple *right rotation* "around node 2" it can be restored again (see Fig. 3.9):

A more complex example:

Inserting 0 into the AVL tree from Fig. 3.10 initially also results in a tree that is not an AVL tree, because the height difference between nodes 4 and 11 is greater than 1. In this case too, a right rotation helps, namely, around node 4, because its left subnode 2 is now left-heavy. However, the node 6 must be "re-hung" because node 9 is now the right subnode

Fig. 3.7 AVL-Baum mit zwei Zahlen

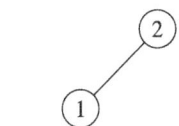

Fig. 3.8 Tree with three numbers

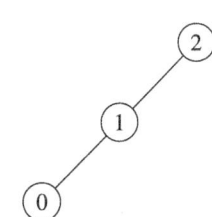

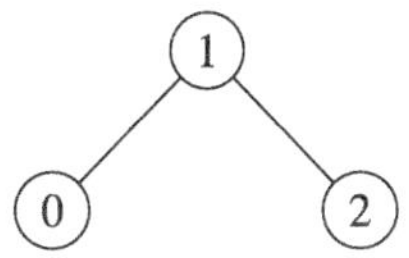

Fig. 3.9 AVL tree with mit three numbers

Fig. 3.10 AVL tree with 11 numbers

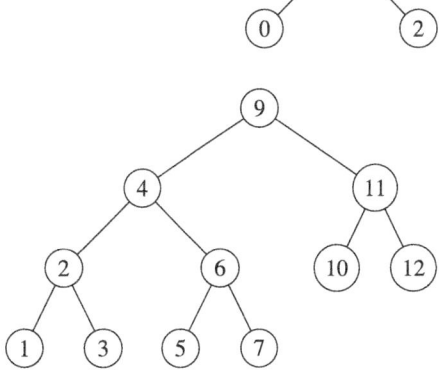

Fig. 3.11 AVL tree with 12 numbers

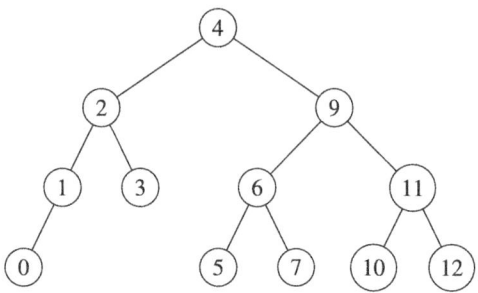

of 4—for this, the pointer to the left subnode of 9 becomes free. So the AVL tree from Fig. 3.11 is created.

The implementation of the right rotation provides a pointer to the left child node y of n which now moves to the position of n and gives its right child node as the left child node to n:

```
// Pre: *x and (*x).left are not empty, *x is leftweighty,
// (*x).left is i) leftweighty or ii) balanced.
//
// i) *x and (*x).right are balanced,
// ii) *x is rightweighty, (*x).right is leftweighty.
func rotR (x *pointer) {
  y := (*x).left
  (*x).left = (*y).right
  (*y).right = *x
  *x = y
  if (*x).balance == leftweighty { // case i)
    (*x).balance = balanced
    (*x).right.balance = balanced
  } else { // case ii)
    (*x).balance = rightweighty
    (*x).right.balance = leftweighty
  }
}
```

In our example, the insertion of 0, case i) applies.

The left rotation rotL() is dual to rotR() in the sense that it arises from rotR() by swapping left and right.

It becomes more difficult when an 8 is inserted into the tree from Fig. 3.10. This initially results in the tree from Fig. 3.12—also not an AVL tree, because as in the previous case the height difference between the nodes 4 and 11 is greater than 1.

A simple right rotation around 4 as in the previous case does not help here, because no AVL tree would result, because then node 4 would have a left subtree of height 2 and a right one of height 4. In this case, the "repair" consists instead of *two* rotations: a left rotation around 4 followed by a right rotation around 6. The result of this (double) left-right rotation around 4 and 6 is the AVL tree from Fig. 3.13.

The implementation of this left-right rotation provides a pointer to the right child node z of the left child node y of n, which has now moved to the position of n, whereby z gives

3.5 Collections of Objects

Fig. 3.12 Baum mit 12 Zahlen

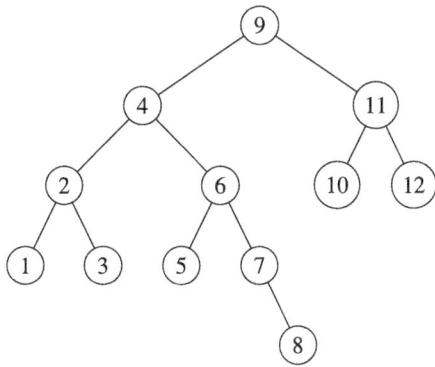

Fig. 3.13 AVL tree with 12 numbers

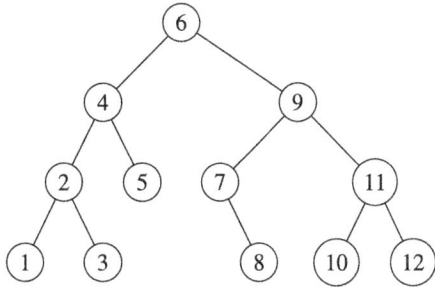

its left child node to y as a right child node, takes over y and gives its right child node to n as a left child node:

```
// Pre: *x, (*x).left and (*x).left.right are not empty,
// (*x) is not balanced,
// (*x) is leftweighty, (*x).left is rightweighty.
// *x is balanced.
func rotLR (x *pointer) {
  y := (*x).left
  z := y.right
  y.right = z.left
  z.left = y
  (*x).left = z.right
  z.right = *x
  *x = z
  switch (*x).balance {
  case leftweighty:
    (*x).left.balance = balanced
    (*x).right.balance = rightweighty
  case balanced:
    (*x).left.balance = balanced
    (*x).right.balance = balanced
  case rightweighty:
    (*x).left.balance = leftweighty
    (*x).right.balance = balanced
  }
  (*x).balance = balanced
}
```

We use the type pointer for pointers to nodes

```
type pointer = *node
```

only to avoid something like *(*node).

The right-left rotation rotRL() is again dual to rotLRL() in the above sense.

We now come to the implementation of the method Ins. If the set is empty, the result is a set of only one element; otherwise

```
func (x *set) Ins (a any) {
  CheckTypeEq (x.any, a)
  if x.anchor == nil {
    x.anchor = newNode (a)
    x.actual = x.anchor
    x.uint = 1
  } else {
    increased := false
    n := ins (&(x.anchor), a, &increased)
    if n != nil {
      x.actual = n
      x.uint++
    }
  }
}
```

The recursive function ins called in the process

```
func ins (x *pointer, a any, increased *bool) pointer {
```

returns the pointer to the inserted node.

The variable increased—initially false—serves the purpose of passing on the information whether the height of a node has increased, each time "one level" further up, so that there—depending on the balance—it can be decided whether a rotation is necessary, and if so, which one.

It is left as an exercise to convince oneself that

- the cases marked with impossible cannot occur and
- in the course of calling an Ins method at most *one* rotation is necessary.

To remove an object from a set, we consider the tree from Fig. 3.14, from which the 10 is to be removed.

The removal of 10 using the method common with binary trees, replacing it with the largest object from the left subtree or the smallest from the right, giving preference to the one of the two possibilities where the subtree has the greater height, leads to the tree in Fig. 3.15.

But in this case, the AVL invariant is violated because the left subtree under 8 has a height that is 2 greater than the right.

In this simple case, we only need a left-right rotation around 3 and 5, which leads to the AVL tree in Fig. 3.16.

This example was simple insofar as only one rotation was needed. With larger trees, restoring the AVL invariant can become considerably more complex, as several rotations

3.5 Collections of Objects

Fig. 3.14 AVL tree with 12 numbers

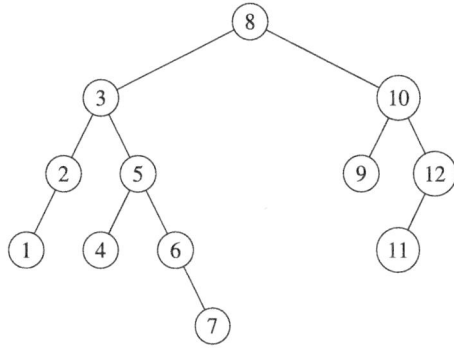

Fig. 3.15 Baum mit 11 Zahlen

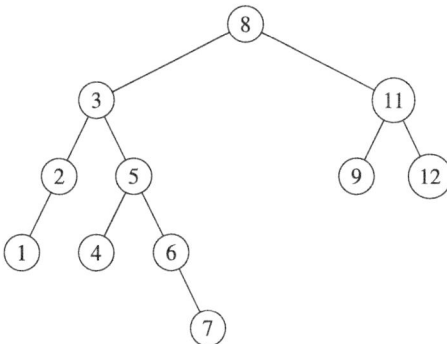

Fig. 3.16 AVL-Baum mit 11 Zahlen

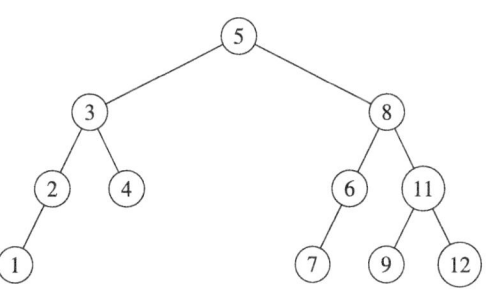

may be necessary and additional measures may need to be taken to balance any resulting imbalances.

We now explain the implementation of the Del method. First, it must be checked whether the object to be removed was the largest in the set. If that was the case, the actual object is now the next smaller one, otherwise the next larger one. After that, the object to be removed is removed using the method common with binary trees.

```
func (x *set) Del() any {
  if x.anchor == nil {
    return nil
  }
  act := x.actual
```

```
toDelete := x.actual.any
x.Step (true) // to set "actual" to the node containing
              // the next largest object, iff such exists
var a any
if act == x.actual { // the node to be deleted is the node with
  a = nil            // the largest object in x, so "actual"
                     // must be set to the node containing
                     // the next smallest object, see below
} else {
  a = Clone (toDelete)
}
decreased := false
if del (&(x.anchor), toDelete, &decreased) { // the object
                          // to be deleted was found and deleted
                          // and the AVL-invariant was secured
  if act == x.actual { // the node to be deleted
                       // was the last right node of x
    if x.uint == 1 {   // see above
      x.actual = nil   // x is now empty
    } else {
      x.Jump (true)    // "actual" is the last right node
    }
  } else { // the node with the next largest object exists
    if x.Ex (a) { // thus the above copy-action to "a":
                  // "actual" might have been rotated off
                  // while deleting, with this trick
                  // it is found again.
    }
  }
  x.uint-
}
return Clone (act.any)
}
```

If this destroys the AVL invariant, it must be restored. This is done with the recursive function `del`:

```
func del (x *pointer, a any, decreased *bool) bool {
  oneLess := false
  if *x == nil {
    return oneLess
  }
  if Less (a, (*x).any) {
    oneLess = del (&((*x).left), a, decreased)
    rebalL (x, decreased)
  } else if Less ((*x).any, a) {
    oneLess = del (&((*x).right), a, decreased)
    rebalR (x, decreased)
  } else { // found node to remove
    if (*x).right == nil {
      *decreased, oneLess = true, true
      *x = (*x).left
    } else if (*x).left == nil {
      *decreased, oneLess = true, true
      *x = (*x).right
    } else if (*x).balance == leftweighty {
      liftL (&((*x).left), *x, decreased, &oneLess)
      rebalL (x, decreased)
    } else {
      liftR (&((*x).right), *x, decreased, &oneLess)
      rebalR (x, decreased)
    }
  }
  return oneLess
}
```

3.5 Collections of Objects

This function uses four other functions: `rebalR` for balancing, if the height difference between two nodes has become greater than 1 and `liftR` for "lifting" a node, if one of its parent nodes was lifted and therefore this place above it has become free, as well as the dual functions `rebalL` and `liftL`.

Here are their implementations:

```
func rebalR (x *pointer, decreased *bool) {
  if *decreased {
    switch (*x).balance {
    case rightweighty:
      (*x).balance = balanced
    case balanced:
      (*x).balance = leftweighty
      *decreased = false
    case leftweighty:
      if (*x).left.balance == rightweighty {
        rotLR (x)
      } else {
        rotR (x)
        if (*x).balance == rightweighty {
              *decreased = false
        }
      }
    }
  }
}
```

and

```
func liftR (x *pointer, y pointer, decreased, oneLess *bool) {
  if (*x).left == nil {
    y.any = Clone ((*x).any)
    *decreased, *oneLess = true, true
    *x = (*x).right
  } else {
    liftR (&((*x).left), y, decreased, oneLess)
    rebalL (x, decreased)
  }
}
```

The following animation program shows (using the undocumented function `Write`) the internal representation of sets in the form of AVL trees: It shows the effects of rotation operations when inserting and removing elements in binary search trees to restore the AVL invariant. It is used (for space reasons) for two-digit numbers.

```
package main
import ("µU/kbd"; "µU/str"; "µU/col"; "µU/mode"
        "µU/scr"; "µU/errh"; "µU/N"; "µU/set")

func main() {
  scr.New (0, 0, mode.XGA); defer scr.Fin()
  cF, cB := col.Black(), col.White()
  scr.ScrColours (cF, cB)
  scr.Cls ()
  scr.Colours (cF, cB)
  N.Colours (cF, cB)
  errh.Hint ("help: F1")
  h := []stringe{"                     Input natural number        ",
                 "and to insert: complete input with Enter",
                 "          remove: complete input with Tab   ",
                 "terminate program: Esc                   "}
  help := make ([]string, len (h))
  for i, c := range (h) { help[i] = str.Lat1 (c) }
```

```
N.SetWd (2)
x := set.New (uint(0))
loop: for {
  scr.Clr (0, 0, scr.NColumns(), scr.NLines() - 1)
  x.Write (0, scr.Wd(), scr.Ht1() / 2, scr.Ht() / 8,
          func (a any) string {
             return N.StringFmt (a.(uint), 2, true)
          })
  N.Colours (cF, cB)
  k := uint(0); N.Edit (&k, 0, 1)
  switch c, _ := kbd.LastCommand(); c {
  case kbd.Esc:
    break loop
  case kbd.Help:
    errh.Help (help)
  case kbd.Enter:
    x.Ins (k)
  case kbd.Tab:
    if x.Ex (k) { x.Del() }
  }
 }
}
```

The implementations of the so far unmentioned methods Empty, Clr, Offc, Num, Jump, Eoc, Get, Put, Trav, Join, Ordered, and Sort are trivial. Surely you can implement these methods without looking into the source codes of the microuniverse.

Also, the recursive methods Ex and ExGeq are—with a view to the recursive structure of the search trees—easy to implement.

It is recommended to implement these methods as an exercise without looking into the microuniverse. The somewhat more complicated method Step remains, but it is not explained here because it is algorithmically uninteresting and rather belongs to the category of "fiddling around".

Why AVL trees are particularly well suited as search trees is clarified by the examination of a special type of AVL trees—the Fibonacci trees—and by some theoretical considerations in the following sections.

3.5.8.1 Fibonacci Trees

Fibonacci trees are recursively defined as follows:

$$F(n) = \begin{cases} \text{empty tree} & \text{für } n < 0 \\ \text{tree with root, } F(n-2) \text{ as left} \\ \text{and } F(n-1) \text{ as right subtree} & \text{for } n \geq 0 \end{cases}$$

The recurrence equation for the number of nodes of these trees in relation to the height n is therefore

$$l(n) = \begin{cases} 1 & \text{für } n = 0 \\ 2 & \text{für } n = 1 \\ 1 + l(n-2) + l(n-1) & \text{für } n \geq 2. \end{cases} \quad (1)$$

3.5 Collections of Objects

These numbers are called Leonardo numbers; they strongly resemble the Fibonacci numbers in their structure.

If we divide the third equation of (1) by $l(n-1)$ and set

$$a_n = \frac{l(n)}{l(n-1)} \quad \text{für } n \geq 1 \tag{2}$$

we get

$$a_n = 1 + \frac{1}{a_{n-1}} + \varepsilon_n \quad \text{with } \varepsilon_n = \frac{1}{l(n-1)}.$$

With a0 = 0 limit formation yields

$$\lim_{n \to \infty} a_n = a, \tag{3}$$

where $a = \frac{1}{2} + \frac{\sqrt{5}}{2} \approx 1{,}618034$ is the positive solution of the quadratic equation $x^2 = 1 + x$ of the golden section.

(2) and (3) provide asymptotically, i.e., for larger n,

$$l(n+1) \approx a \cdot l(n),$$

therefore due to $l(0) = 1$ for the node number $l(n)$ depending on the height n asymptotically

$$l(n) \approx a^n. \tag{4}$$

For the inversion, i.e., the calculation of the height n for a given node number k, we obtain by logarithmizing

$$\log_2 k \approx \log_2(a^n) = n \cdot \log_2 a,$$

consequently

$$n \approx \frac{\log_2 k}{\log_2 a} = 1{,}44 \cdot \log_2 k$$

with $1{,}44 \approx \frac{1}{\log_2 1{,}618034}$.

By induction, it is easy to prove that Fibonacci trees

- are AVL trees (because $F(n)$ has the height n) and that they
- have a minimum number of nodes for a given height (if a node is removed in the left or right subtree, the height of the tree decreases by 1—in the first case after a simple left rotation).

They are therefore in *this sense* the "worst possible" AVL trees, in that every AVL tree of a given height has at least as many nodes as the Fibonacci tree of the same height—in other words, an AVL tree cannot be taller than a Fibonacci tree with the same number of nodes.

According to the considerations from the previous section, *the height of an AVL tree in the worst case is only about 44% greater than that of a best possible balanced tree with the same number of nodes*, i.e., in particular, logarithmic search time is guaranteed with them.

3.5.8.2 Explicit Representation of the Leonardo Numbers

The Leonardo numbers (*number of nodes in Fibonacci trees*) are recursively defined by

$$l(n) = \begin{cases} 0 & \text{for } n \leq 0 \\ 1 + l(n-2) + l(n-1) & \text{for } n > 0 \end{cases}. \tag{1}$$

For $n \leq 0$ the sequence

$$0, 1, 2, 4, 7, 12, 20, 33, 54, 88, 143, 232, 376, \ldots$$

results.

By using two accumulators, this can also be formulated end-recursively:

$$l(0) = 0$$
$$l(n) = l'(n-1, 1, 0) \quad \text{für } n > 0$$
$$\text{with } l'(n, a, b) = \begin{cases} a & \text{for } n = 0 \\ l'(n-1, 1+a+b, a) & \text{for } n > 0. \end{cases}$$

With the formal power series (see Sect. 3.5.8.3)

$$f(X) = \sum_{n=0}^{\infty} l(n) X^n = X + 2X^2 + 4X^3 + 7X^4 + 12X^5 + \ldots \tag{2}$$

and

$$q(X) = \frac{1}{1-X} = \sum_{n=0}^{\infty} X^n = 1 + X + X^2 + X^3 + X^4 + \ldots$$

we obtain by substituting (1)

$$Xf(X) + X^2 f(X) + q(X) = \sum_{n=0}^{\infty} l(n) X^{n+1} + \sum_{n=0}^{\infty} l(n) X^{n+2} + \sum_{n=0}^{\infty} X^n$$

$$= \sum_{n=0}^{\infty} \big(l(n-1) + l(n-2) + 1\big) X^n$$

$$= 1 + \sum_{n=0}^{\infty} l(n) X^n = 1 + f(X).$$

3.5 Collections of Objects

resolved for $f(X)$:
$$f(X) = \frac{X}{(1-X-X^2)(1-X)}. \tag{3}$$

The factorization of $1 - X - X^2 = (1-aX)(1-bX)$ yields the system of equations

$$a + b = 1 \tag{3.1}$$

$$ab = -1 \tag{4}$$

with the solutions
$$a = \frac{1}{2} + \frac{\sqrt{5}}{2}, \quad b = \frac{1}{2} - \frac{\sqrt{5}}{2}. \tag{5}$$

The approach to the partial fraction decomposition of (3)

$$\frac{1}{(1-X-X^2)(1-X)} = \frac{A}{1-aX} + \frac{B}{1-bX} + \frac{C}{1-X} \tag{6}$$

yields the equation

$$A(1-bX)(1-X) + B(1-aX)(1-X) + C(1-aX)(1-bX) = 1,$$

and from this with (4) by comparing coefficients the linear system of equation (7)

$$\begin{aligned} A+B+C &= 1 \\ (1+b)A + (1+a)B + C &= 0 \\ bA + aB - C &= 0 \\ A+B+C &= 1 \end{aligned} \tag{7}$$

with the solutions

$$A = 1 + \frac{2}{5}\sqrt{5}, \quad B = 1 - \frac{2}{5}\sqrt{5}, \quad C = -1. \tag{8}$$

Substituting (6) into (3) in conjunction with the general geometric power series

$$\frac{1}{1-cX} = \sum_{n=0}^{\infty} c^n X^n$$

gives

$$f(X) = \frac{A}{1-aX}X + \frac{B}{1-bX}X + \frac{C}{1-X}X$$
$$= A\sum_{n=0}^{\infty} a^n X^{n+1} + B\sum_{n=0}^{\infty} b^n X^{n+1} + C\sum_{n=0}^{\infty} X^{n+1}$$
$$= \sum_{n=0}^{\infty}(Aa^n + Bb^n + C)X^{n+1}$$
$$= \sum_{n=1}^{\infty}(Aa^{n-1} + Bb^{n-1} + C)X^n.$$

After (2) we obtain by comparing coefficients for the Leonardo numbers

$$l(n) = Aa^{n-1} + Bb^{n-1} - 1 \quad \text{für } n > 0.$$

The third equation of the system (7) yields $\frac{A}{a} + \frac{B}{b} - 1 = 0$, therefore this result is also correct for $n = 0$.

By substituting (5) and (8), we obtain the explicit representation of the Leonardo numbers

$$l(n) = \left(1 + \frac{2}{5}\sqrt{5}\right)\left(\frac{1}{2} + \frac{\sqrt{5}}{2}\right)^{n-1} + \left(1 - \frac{2}{5}\sqrt{5}\right)\left(\frac{1}{2} - \frac{\sqrt{5}}{2}\right)^{n-1} - 1$$

for all natural numbers n.

3.5.8.3 Formal Power Series

The manipulations of the power series in the previous section are justified by the following considerations.

For sets A and B, let A^B denote the set of mappings from B to A. Let X be an indeterminate (a symbol).

We consider the set $M = \mathbb{N}^2\{X\} = \mathbb{N}\{X\}$. It holds $H \approx \mathbb{N}$, because the mapping

$$f: \mathbb{N} \to H, \quad \text{defined by} \quad f(n)(X) = n,$$

is bijective; with the notation $X^n = f(n) \in H$ is

$$H = \{X^n \mid n \in \mathbb{N}\}.$$

H forms with respect to the multiplication defined by

$$X^n \cdot X^k = X^{n+k}$$

3.5 Collections of Objects

with $1 = X^0$ as the neutral element a commutative semigroup, which can be easily verified by calculation; f is furthermore an isomorphism due to $f(0) = X^0 = 1$ and $f(n+k) = X^{n+k} = X^n \cdot X^k = f(n) \cdot f(k)$.

For a commutative ring A, let

$$A[[X]] = A^H.$$

Every $p \in A[[X]]$ can then be uniquely written in the form

$$p = \sum_{n \in \mathbb{N}} a_n X^n = a_0 + a_1 X + a_2 X^2 + \ldots \quad \text{mit} \quad a_n = p(X^n)$$

because p can be uniquely recovered from the term $\sum_{n \in \mathbb{N}} a_n X^n$ by the definition $p(X^n) = a_n$. By

$$\sum_{n \in \mathbb{N}} a_n X^n + \sum_{n \in \mathbb{N}} b_n X^n = \sum_{n \in \mathbb{N}} c_n X^n \quad \text{mit} \quad c_n = a_n + b_n$$

und $\sum_{n \in \mathbb{N}} a_n X^n \cdot \sum_{n \in \mathbb{N}} b_n X^n = \sum_{n \in \mathbb{N}} c_n X^n \quad \text{mit} \quad c_n = \sum_l imitsi, k \in \mathbb{N}, i + k = n a_i b_k$

an addition and a multiplication on $A[[X]]$ are defined, with respect to which $A[[X]]$ forms a commutative ring (proof by calculation).

$A[[X]]$ is called the *ring of formal power series over A*.

By induction, it can be shown that $A[[x]]$ is free of zero divisors if A is; for a field A, the *quotient ring of formal power series* (the ring of fractions of formal power series) is therefore also a *field*, i.e., one can calculate with these fractions, e.g., over $\mathbb{Q}$ or $\mathbb{R}$—just like with numbers.

Finally, it should be noted that this construction includes that of a polynomial ring $A[X]$ in an indeterminate X over a ring A:

Replace $H = \mathbb{N}^{\{X\}}$ with $\mathbb{N}^{(\{X\})}$ where $A^{(B)}$ denotes for a ring A the set of mappings from B to A that are almost everywhere 0, i.e., for which $f(b) \neq 0$ only holds for finitely many $b \in B$.

The fractionally rational functions over $\mathbb{R}$ thus form a subfield of the quotient field of the formal power series $R[[X]]$.

3.5.9 Persistent Sequences (Sequential Files)

Under *persistent objects*, we understand such objects to be that are *permanently* stored on a data carrier, i.e., when a program that uses them is called, they have *those* values that they had at the end of the last program run.

Their specification differs from that of the *collections* only *in that* they have a "*handle*" with which they can be *found again*: a uniquely identifiable *name* in the file system.

For this, we need the following interface, which is also located in the file `persistor.go` in the package `obj`:

```
package obj

type Persistor interface {
// An object "is defined with a name" means, that it is stored
// in a persistent file with that name as "handle" in the filesystem.

// Pre: n is a valid name in the filesystem and there exists no object
// of a type different from the type of x, but defined with name n.
// x is now defined with name n, i.e. it is the object, that is stored
// in a file with that name, if there exists such; otherwise it is
   empty.
  Name (n string)

// x is defined with that name.
// Another file with that name is now destroyed.
  Rename (n string)

// Pre: x is defined with a name.
// x is secured in the file system.
  Fin ()
}

func IsPersistor (a any) bool {
  if a == nil { return false }
  _, ok := a.(Persistor)
  return ok
}
```

Persistent sequences are thus *sequential files*.

The microuniverse contains for persistent sequences of atomic variables or objects of type `Object` the package `pseq`, which provides the corresponding abstract data type:

```
package pseq
import . "µU/obj"

type PersistentSequence interface {

  Seeker // hence Collector
  Persistor
}

// Pre: a is atomic or of a type implementing Equaler and Coder.
// Returns a new empty persistent sequence for objects of the type of a
 .
func New (a any) PersistentSequence { return new_(a) }

func Length (s string) uint { return length(s) }
```

Their *implementation* consists of sequential files of byte sequences of the codelength of the objects. The length of such a file is the product of the codelength of an object and the number of objects.

The representation of a persistent sequence

```
type persistentSequence struct {
  name, tmpName string
  ordered bool
  any "pattern object"
  object any
  file internal.File
  owner, group uint
```

3.5 Collections of Objects

```
  size, pos, num uint64
  buf, buf1 Stream
}
```

requires the internal abstract data type

```
package internal
import . "µU/obj"

type File interface {
  Fin()
  Name (n string)
  Rename (n string)
  Empty() bool
  Clr()
  Length() uint64
  Seek (p uint64)
  Position() uint64
  Read (s Stream) (int, error)
  Write (s Stream) (int, error)
}
func DirectLength (n string) uint64 { return directLength(n) }
func Erase (n string) { erase(n) }

func New() File { return new_() }
```

which encapsulates the accesses into the file system.

Behind the implementation of this data type, the classic concept of accessing sequential files shines through, as realized by Niklaus Wirth in *Pascal* (with a view to the historical situation: tapes as peripheral data carriers) with the operations Reset, Rewrite, Read, Write, Eof, and Seek.

The implementations of the methods are not explained here because there are no interesting algorithms behind them. For the same reasons as with the sequences (see end of Sect. 3.5.4, many algorithms here also have linear complexity.

3.5.10 Persistent Index Sets

Consistently, this section should deal with *persistent sets*. For this, there is a good data type: that of the *B-trees* of Bayer and McCreight (see [4]).

Instead, we present a concept here that carries much further: the *persistent index sets*. These are persistent sets, on whose elements can be accessed directly via a key.

We require that the objects in the persistent sets implement the type Object and that the keys either have an atomic type with an order relation or implement the types Equaler and Comparer.

The function that assigns a key to an object must be injective, i.e., different objects must have different keys, so that the search for an object via a key provides a unique result. We call this "key function" Index (hence the name for the data type).

Objects that have such keys, we call *Indexer*. They implement the interface

```
package obj

type Indexer interface {

  Object
  Editor

  Index() Func
}
func IsIndexer (a any) bool {
  if a == nil { return false }
  _, ok := a.(Indexer)
  return ok
}
```

in the file `indexer.go` in the package `obj`.

Here are some examples: The index

- of a person in an address directory with various information is the pair consisting of the person's name and date of birth (where, if necessary, identical names must be differentiated by nicknames);
- of a book in a book directory is the tuple of title and author[s]; and
- of a page in a diary is the calendar date of the page.

The microuniverse contains for the persistent ordered sets of objects of type `Object` with index the package `piset`, defined by the interface:

```
package piset
import . "µU/obj"

type PersistentIndexedSet interface { // persistent ordered sets
                                      // of objects, that have an index
                                      // by which they are ordered.
  Collector
  Persistor
  Operate()
}
// Returns a new empty persistent indexed set
// for objects of the type of o.
func New (o Indexer) PersistentIndexedSet { return new_(o) }
```

These sequential files (of constant size byte sequences) are equipped with index trees and a position management system.

The basic idea in the implementation is the management of the keys to the objects together with their positions in the file in a (non-persistent) set `set.Set`. This has a major advantage:

- The position of an object can be quickly found via its key (see last sentence in Sect. 3.5.8.1), where the search takes place in memory,
- and thus direct access to the object in the file is possible.

The pairs of key and associated file positions form a data type that is housed in a subpackage of `piset`:

3.5 Collections of Objects

```
package pair
import . "µU/obj"

type Pair interface {

  Equaler
  Comparer
  Pos() uint
  Index() any
}

func New (a any, n uint) Pair { return new_(a,n) }
```

The representation of such a pair and the implementation of the constructor is simple:

```
package pair
import . "µU/obj"

type pair struct {
                   any  "index"
                   uint "position"
                   }
func new_(a any, n uint) Pair {
  x := new(pair)
  x.any = Clone(a)
  x.uint = n
  return x
}
```

With this, we have the representation of the data type:

```
type persistentIndexedSet struct {
                                   Object "pattern object"
                                   any "index of Object"
                                   pseq.PersistentSequence "file"
                                   Func "index function"
                                   set.Set "pairs of index and position
                                        in the file"
                                   buf.Buffer "free positions in the file
                                        "
}
```

The type Func stands for

```
type Func func (any) any
```

value-delivering functions with one argument.

The constructor is implemented as follows:

```
func new_(o Object, f Func) PersistentIndexedSet {
  x := new (persistentIndexedSet)
  x.Object = o.Clone().(Object)
  x.any = f(o)
  x.PersistentSequence = pseq.New (x.Object)
  x.Func = f
  x.Set = set.New (pair (x.any, 0))
  x.Buffer = buf.New (uint(0))
  return x
}
```

During the course of a program in which piset is imported, the underlying file pseq.PersistentSequence, if it is not empty, is traversed at the beginning, with the following happening:

- If the object at the relevant position in the file is empty, the actual position is inserted into the buffer of free positions;
- otherwise, by applying the key function Func passed in the constructor, the key is calculated and inserted into the key set set.Set together with the current position.

This happens when identifying the file by its name:
```
func (x *persistentIndexedSet) Name (s string) {
  if str.Empty (s) { return }
  x.PersistentSequence.Name (s + ".seq")
  x.Set.Clr()
  x.Buffer = buf.New (uint(0))
  if x.PersistentSequence.Empty() { return }
  for i := uint(0); i < x.PersistentSequence.Num(); i++ {
    x.PersistentSequence.Seek (i)
    x.Object = x.PersistentSequence.Get().(Object)
    if x.Object.Empty() {
      x.Buffer.Ins (i)
    } else {
      x.Set.Ins (internal.New (x.Func (x.Object), i))
    }
  }
  x.Jump (false)
}
```

The disadvantage that this delay brings with it is more than offset by the aforementioned advantage.

An object is *inserted* into the file by inserting its key and its position in the file into set.Set. If the buffer buf.Buffer of free positions is not empty, the object encoded as a byte sequence is written into the file from the position read from the buffer, otherwise it is appended to the end of the file.

Here is the implementation of the method Ins:
```
func (x *persistentIndexedSet) Ins (a any) {
  x.check (a)
  if x.Ex (a) || a.(Object).Empty() { return }
  var n uint
  if x.Buffer.Empty() {
    n = x.PersistentSequence.Num()
  } else {
    n = x.Buffer.Get().(uint)
  }
  x.PersistentSequence.Seek (n)
  x.PersistentSequence.Put (a)
  x.Set.Ins (pair (x.Func (a), n))
  x.PersistentSequence.Seek (n)
}
```

An object is *removed* by fetching (and thereby removing) the pair of its key and its position in the file from set.Set, then a byte sequence representing an empty object is placed from its file position, and finally this position is inserted into the buffer.

Here is the implementation of the Del method:
```
func (x *persistentIndexedSet) Del() any {
  if x.Set.Empty() {
    x.Object.Clr()
    return x.Object
  }
```

3.5 Collections of Objects

```
  n := x.Set.Get().(internal.Pair).Pos()
  x.PersistentSequence.Seek (n)
  x.Object = x.PersistentSequence.Get().(Object)
  object := x.Object.Clone().(Object)
  object.Clr()
  x.PersistentSequence.Put (object)
  x.Buffer.Ins (n)
  if ! x.Set.Empty() {
    n := x.Set.Get().(internal.Pair).Pos()
    x.PersistentSequence.Seek (n)
  }
  return x.Object.Clone()
}
```

and here is the implementation of the Get method for reading the actual object:

```
func (x *persistentIndexedSet) Get() any {
  if x.Set.Empty() {
    x.Object.Clr()
    return x.Object
  }
  p := x.Set.Get().(internal.Pair)
  n := p.Pos()
  x.PersistentSequence.Seek (n)
  return x.PersistentSequence.Get().(Object)
}
```

The implementation of the other methods is trivial; they ultimately always rely on the same principle of directly accessing the objects in the file via the key set set.Set.

3.5.11 Graphs

Graphs play an important role for many applications; typical examples are

- subway or bus networks with stations or stops as nodes and
- the connections between them as edges.

or

- maps with cities as nodes and
- the roads between them as edges.

Figure 3.17 shows an example.

The microuniverse contains in the package gra the abstract data type Graph, *graphs* of objects of atomic data types or of type Object

The number of nodes in graphs can be arbitrarily large—within the available memory resources—the number of edges is also not limited.

The topic "*graph algorithms*" is so comprehensive that it fills books.

Some of them are implemented in the graph package, e.g.:

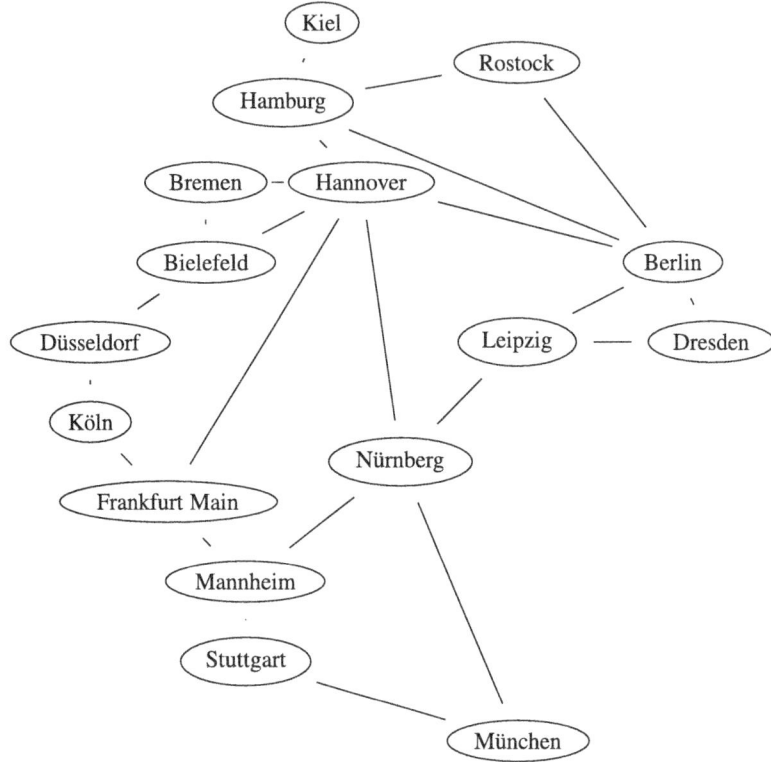

Fig. 3.17 Example of a graph

- for searching for the shortest connection between two nodes (main purpose of navigation devices);
- for searching for circles (chains of connections with the same starting and ending node), thus also for investigating whether there are any circles in a graph at all;
- for searching for *Euler paths* (chains of connections that reach each node exactly once);
- for searching for minimal *spanning trees* indexspanning tree (subgraphs from all corners in the form of a tree).

The package is used in the projects:

- Robi, the robot (Chap. 5).
- Railway (Chap. 14).
- Berlin U- and S-Bahn (Chap. 16).

The specification is quite long for obvious reasons:

3.5 Collections of Objects

```
package gra
import ( . "µU/obj" "µU/adj" "µU/pseq")

type Demo byte // for demonstration purposes

const (Depth = Demo(iota); Cycle; Euler; TopSort;
       ConnComp; Breadth; SpanTree; nDemos)

type Demoset [nDemos]bool

// Sets of vertices with an irreflexive relation:
// Two vertices are related, iff they are connected by an edge, where
//   there
// are no loops (i.e. no vertex is connected with itself by an edge).
// If the relation is symmetric, the graph is called "undirected",
// if it is strict, "directed" (i.e. all edges have a direction).
//
// The edges have a number of type uint as value ("weight");
// either all edges have the value 1 or their value is given by
// the function Val (they have to be of an uint-type or of type
//   Valuator).
// The outgoing edges of a vertex are enumerated (starting with 0);
// the vertex, with which a vertex is connected by its n-th outgoing
//   edge,
// is denoted as its n-th neighbourvertex.
//
// In any graph some vertices and edges might be marked.
//
// A path in a graph is a sequence of vertices and from each of those
// - excluding from the last one - an outgoing edge to the next vertex.
// A simple path is a path of pairwise disjoint vertices.
// An Euler path is a path that traverses each edge exactly once
// (it may pass any vertex more than once).
// A cycle is a path with an additional edge
// from the last vertex of the path to its first.
//
// A graph G is (strongly) connected, if for any two vertices
// v, v1 of G there is a path from v to v1 or (and) vice versa;
// so for undirected graphs this is the same.
//
// In any nonempty graph exactly one vertex is distinguished as colocal
// and exactly one as local vertex.
// Each graph has an actual path.

type Graph interface {

  Object

  Persistor

// Returns true, iff x is directed.
  Directed() bool

  SetDir (b bool)

// Returns a copy of the graph that is not directed.
  Indir() Graph

// Returns the number of vertices of x.
  Num() uint

// Returns the number of edges of x.
  Num1() uint

// Returns the number of marked vertices of x.
  NumMarked() uint
```

```
// Returns the number of marked edges of x.
  NumMarked1() uint

// Pre: p is defined on vertices.
// Returns the number of vertices of x, for which p returns true.
  NumPred (p Pred) uint

// If v is not of the vertextype of x or if v is already contained
// as vertex in x, nothing has happend. Otherwise:
// v is inserted as vertex in x.
// If x was empty, then v is now the colocal and local vertex of x,
// otherwise, v is now the local vertex and the former local vertex
// is now the colocal vertex of x.
  Ins (v any)

// If x was empty or if the colocal vertex of x coincides
// with the local vertex of x or if e is not of the edgetype of x,
// nothing has happened. Otherwise:
// e is inserted into x as edge from the colocal to the local vertex of
    x
// (if these two vertices were already connected by an edge,
// that edge is replaced by e).
// For e == nil e is replaced by uint(1).
  Edge (e any)

// If x is empty or has an edgetype or
// if v or v1 is not of the vertextype of x or
// if v or v1 is not contained in x or
// if v and v1 coincide or
// if a is not of the type of the pattern edge of x or
// if there is already an edge from v to v1,
// nothing has happened. Otherwise:
// v is now the colocal and v1 the local vertex of x
// and e is inserted is an edge from v to v1.
  Edge2 (v, v1, e any)

// Returns the representation of x as adjacency matrix.
  Matrix() adj.AdjacencyMatrix

// Pre: m is symmetric iff x is directed.
// x is the graph with the vertices a.Vertex(i) and edges from
// a.Vertex(i) to a.Vertex(k), iff a.Val(i,k) > 0 (i, k < a.Num()).
  SetMatrix (a adj.AdjacencyMatrix)

// Returns true, iff the colocal vertex of x does not
// coincide with the local vertex of x and there is
// an edge in x from the colocal to the local vertex.
  Edged() bool

// Returns true, iff
// the colocal vertex does not coincide with the local vertex of x
// and there is an edge from the local to the colocal vertex in x.
  CoEdged() bool

// Returns true, iff v is contained as vertex in x.
// In this case, v is now the local vertex of x.
// The colocal vertex of x is the same as before.
  Ex (v any) bool

// Returns true, if v and v1 are contained as vertices in x
// and do not coincide. In this case now
// v is the colocal and v1 the local vertex of x.
  Ex2 (v, v1 any) bool

// Pre: p is defined on vertices.
// Returns true, iff there is a vertex in x, for which p returns true.
// In this case some now such vertex is the local vertex of x.
```

3.5 Collections of Objects

```
// The colocal vertex of x is the same as before.
   ExPred (p Pred) bool

/// Returns true, iff e is contained as edge in x.
// In this case the neighbour vertices of some such edge are now
// the colocal and the local vertex of x (if x is directed,
// the vertex, from which the edge goes out, is the colocal vertex.
   Ex1 (e any) bool

// Pre: p is defined on edges.
// Returns true, iff there is an edge in x, for which p returns true.
// In this case the neighbour vertices of some such edge are now
// the colocal and the local vertex of x (if x is directed,
// the vertex, from which the edge goes out, is the colocal vertex.
   ExPred1 (p Pred) bool

// Pre: p and p1 are defined on vertices.
// Returns true,
// iff there are two different vertices v and v1 with p(v) and p(v1).
// In this case now some vertex v with p(v) is the colocal vertex
// and some vertex v1 with p1(v1) is the local vertex of x.
   ExPred2 (p, p1 Pred) bool

// Returns the value of the local vertex of x,
// if it has the type Valuator; return otherwise 1.
// Val() uint)

// Returns true, iff x contains a vertex with the value n.
// In this case, such a vertex is the local vertex of x.
// The colocal vertex of x is the same as before.
   ExVal (n uint) bool

// Returns true, iff x contains a vertex v with the value n
// and a vertex v1 with the value n1. In this case,
// v is the colocal vertex of x and v1 is the local vertex of x.
   ExVal2 (n, n1 uint) bool

// Returns the pattern vertex of x, if x is empty;
// returns otherwise a clone of the local vertex of x.
   Get() any

// Returns a clone of the pattern edge of x, if x is empty
// or if there is no edge from the colocal vertex to the
// local vertex of x or if these two vertices coincide.
// Returns otherwise a clone of the edge from the
// colocal vertex of x to the local vertex of x.
   Get1() any

// Returns (nil, nil), if x is empty.
// Returns otherwise a pair, consisting of clones
// of the colocal and of the local vertex of x.
   Get2() (any, any)

// If x is empty or if v is not of the vertex type of x, nothing has
   happened. Otherwise:
// The local vertex of x is replaced by v.
   Put (v any)

// If x is empty or if e has no edge type or
// if e is not of the edgetype of x or
// if there is no edge from the colocal to the local vertex of x,
// nothing has happened. Otherwise:
// The edge from the colocal to the local vertex of x is replaced by e.
   Put1 (e any)

// If x is empty or if v or v1 is not of the vertextype of x or
// if the colocal vertex of x coincides with the local vertex,
```

```
// nothing had happened. Otherwise:
// The colocal vertex of x is replaced by v
// and the local vertex of x is replaced by v1.
   Put2 (v, v1 any)

// No vertex and no edge in x is marked.
   ClrMarked()

// If x is empty or if v is not of the vertex type of x
// or if v is not contained in x, nothing has happened.
// Otherwise, v is now the local vertex of x and is marked.
// The colocal vertex of x is the same as before.
   Mark (v any)

// If x is empty or if v or v1 is not of the vertex type of x
// or if v or v1 is not contained in x
// or if v and v1 conincide, nothing had happened.
// Otherwise, v is now the colocal and v1 the local vertex
// of x and these two vertices and the edge between them are now marked
   .
   Mark2 (v, v1 any)

// Returns true, if all vertices and all edges of x are marked.
   AllMarked() bool

// If x is empty, nothing has happened. Otherwise:
// The former local vertex of x and
// all its outgoing and incoming edges are deleted.
// If x is now not empty, some other vertex is now the local vertex
// and coincides with the colocal vertex of x.
// The actual path is empty.
   Del()

// If there was an edge between the colocal and the local vertex of x,
// it is now deleted from x.
   Del1()

// Returns true, iff x is empty or
// if the colocal vertex coincides with the local vertex of x or
// if there is a path from the colocal to the local vertex in x.
   Conn() bool

// Pre: p is defined on vertices.
// Returns true, iff x is empty or
// the colocal vertex coincides with the local vertex of x or
// if p returns true for the local vertex and there is a path
// from the colocal vertex of x to the local vertex, that contains
// - apart from the colocal vertex - only vertices, for which p returns
     true.
   ConnCond (p Pred) bool

// If x is empty, nothing had happened. Otherwise:
// If there is a path from the colocal to the local vertex of x,
// the actual path of x is a shortest such path
// (shortest w.r.t. the sum of the values of its edges,
// hence, if x has no edgetype, w.r.t. their number).
// If there is no path from the colocal to the local vertex of x,
// the actual path consists only of the colocal vertex.
// The marked vertices and edges of x are
// the vertices and edges in the actual path of x.
   FindShortestPath()

// Pre: p is defined on vertices.
// If x is empty, nothing had happened. Otherwise:
// If p returns true for the local vertex and there is a path
// from the colocal to the local vertex of x, that contains
```

3.5 Collections of Objects

```
// - apart from the colocal vertex - only vertices, for which p returns
     true,
// the actual path of x is a shortest such path
// w.r.t. the sum of the values of its edges
// (hence, if x has no edgetype, w.r.t. their number).
// Otherwise the actual path consists only of the colocal vertex.
// The marked vertices and edges of x are
// the vertices and edges in the actual path of x.
  FindShortestPathPred (p Pred)

// Pre: Act or ActPred was called before.
// Returns the slice of the vertices of the actual path.
  ShortestPath() []any

// Returns the sum of the values of all edges of x
// (hence, if x has no edgetype, the number of the edges of x).
  Len() uint

// Returns the sum of the values of all marked edges in x
// (hence, if x has no edgetype, the number of the marked edges).
  LenMarked() uint

// Returns 0, if x is empty.
// Returns otherwise the number of the outgoing edges of the local
     vertex of x.
  NumNeighboursOut() uint

// Pre: x is directed.
// Returns 0, if x is empty.
// Returns otherwise the number of the incoming edges to the local
     vertex of x.
  NumNeighboursIn() uint

// Returns 0, if x is empty.
// Returns otherwise the number of all edges of the local vertex of x.
  NumNeighbours() uint

// If x is not directed, nothing had happened. Otherwise:
// The directions of all edges of x are reversed.
  Inv()

// If x is not directed, nothing had happened. Otherwise:
// The directions of all outgoing and incoming edges
// of the local vertex of x are reversed.
  InvLoc()

// If x is empty, nothing had happened. Otherwise:
// The local and the colocal vertex of x are exchanged.
// The actual path of x consists only of the colocal vertex of x.
// The only marked is the colocal vertex; no edges are marked.
  Relocate()

// If x is empty, nothing had happened. Otherwise:
// The colocal vertex of x coincides with the local vertex of x,
// where for f == true that is the vertex, that was the former local
     vertex of x,
// and for !f the vertex, that was the former colocal vertex of x.
// The actual path of x consists only of this vertex.
// The only marked vertex is this vertex; no edges are marked.
  Locate (f bool)

// Returns true, iff x is empty or the local vertex of x
// coincides with the colocal vertex of x.
  Located() bool

// If x is empty, nothing had happened. Otherwise:
// The local and the colocal vertex of x are exchanged;
```

```
// the actual path is not changed and
// the marked vertices and edges are unaffected.
   Colocate()

// If x is empty or directed, nothing has happened.
// Otherwise the actual path of x is inverted, particularly
// the local and the colocal vertex of x are exchanged.
// The marked vertices and edges are unaffected.
   InvertPath()

// If x is empty or if i >= number of vertices outgoing from the local
   vertex
// nothing had happened. Otherwise:
// For f:  The i-th neighbour vertex of the last vertex of the actual
   path
//         of x is appended to it as new last vertex.
// For !f: The last vertex of the actual path of x is deleted from it,
//         if it had not only one vertex (i does not play any role in
   this case).
// The last vertex of the actual path of x is the local vertex of x and
// Vertices and edges in x are marked, if the belong to its actual path
   .
   Step (i uint, f bool)

// Returns false, if x is empty or if i >= NumNeighbours();
// returns otherwise true, iff the edge to the i-th neighbour
// of the local vertex is an outgoing edge.
   Outgoing (i uint) bool

// Returns nil, if x is empty or if i >= NumNeighboursOut();
// returns otherwise a clone of the i-th outgoing neighbour of the
   local vertex.
   NeighbourOut (i uint) any

// Returns false, if x is empty or if i >= NumNeighbours();
// returns otherwise true, iff the edge to the i-th neighbour
// of the local vertex is an incoming one.
   Incoming (i uint) bool

// Returns nil, if x is empty or if i >= NumNeighboursIn();
// returns otherwise a copy of the its i-th incoming neighbour of the
   local vertex.
   NeighbourIn (i uint) any

// Returns nil, if x is empty or if i >= NumNeighbours();
// returns otherwise a clone of its i-th neighbour vertex
// of the local vertex of x.
   Neighbour (i uint) any

// Pre: p is defined on vertices.
// Returns true, if x is empty or
// if p returns true for all vertices of x.
   True (p Pred) bool

// Pre: p is defined on vertices.
// Returns true, iff x is empty or
// if p returns true for all marked vertices in x.
   TrueMarked (p Pred) bool

// Pre: o is defined on vertices.
// o is applied to all vertices of x.
// The colocal and the local vertex of x are the same as before;
// the marked vertices and edges are unaffected.
   Trav (o Op)

// Pre: o is defined on vertices.
// o is applied to all vertices of x, where
```

3.5 Collections of Objects

```
// o is called with 2nd parameter "true", iff
// the corresponding vertex is marked.
// Colocal and local vertex of x are the same as before;
// The marked edges are unaffected.
  TravCond (o CondOp)

// Pre: o is defined on edges.
// If x has no edgetype, nothing had happened. Otherwise:
// o is applied to all edges of x.
// Colocal and local vertex of x are the same as before;
// the marked vertices and edges are unaffected.
  Trav1 (o Op)

// Pre: o is defined on edges.
// If x has no edgetype, nothing had happened. Otherwise:
// o is applied to all edges of x with 2nd parameter "true",
// iff the correspoding edge is marked.
// Colocal and local vertex of x are the same as before;
// the marked vertices and edges are unaffected.
  Trav1Cond (o CondOp)

// Pre: o is defined on edges.
// If x has no edgetype, nothing had happened. Otherwise:
// o is applied to all edges of the local vertex of x.
  Trav1Loc (o Op)

// Pre: o is defined on edges.
// If x has no edgetype, nothing had happened. Otherwise:
// o is applied to all edges of the colocal vertex of x.
  Trav1Coloc (o Op)

// Returns nil, if x is empty.
// Returns otherwise the graph consisting of the local
// vertex of x, all its neighbour vertices and of all edges
// outgoing from it and incoming to it.
// The local vertex of x is the local vertex of the star.
// It is the only marked vertex in the star;
// all edges in the star are marked.
  Star() Graph

// Returns true, iff there are no cycles in x.
  Acyclic() bool

// If x is empty, nothing has happened. Otherwise:
// The following equivalence relation is defined on x:
// Two vertices v and v1 of x are equivalent, iff there is
// a path in x from v to v1 and vice versa (hence the set of
// equivalence classes is a directed graph without cycles).
  Isolate() // TODO name

// Exactly those vertices in x are marked, that are equivalent
// to the local vertex and of exactly all edges between them.
// No edges in x are marked.
  IsolateMarked() // TODO name

// Returns true, iff x is not empty and
// if the local and the colocal vertex of x are equivalent,
// i.e. for both of them there is a path in x to the other one.
  Equiv() bool

// Returns false, if x is not totally connected.
// Returns otherwise true, iff there is an Euler path or cycle in x.
  Euler() bool

// If x is directed, nothing has happened. Otherwise:
// Exactly those vertices and edges in x are marked,
// that build a minimal spanning tree in the connected component
```

```
// containing the colocal vertex
// (minimal w.r.t. the values of the sum of its edges;
// hence, if x has no edgetype, w.r.t. the number of its vertices)
// The actual path is not changed.
  MST()

// If x is empty or undirected or
// if x is directed and has cycles, nothing has happened. Otherwise:
// The vertices of x are ordered s.t. at each subsequent traversal of x
// each vertex with outgoing edges is always handled before the
    vertices,
// at which those edges come in.
  Sort()

// Pre: x is directed, iff all graphs y are directed.
// x consists of all vertices and edges of x before
// and of all graphs y. Thereby all marks of y are overtaken.
  Add (y ...Graph)

// The demofunction for d is switched on, iff s[d] == true.
  SetDemo (d Demo)

// Pre: wv is defined on vertices and we on edges.
// wv and we are the actual write functions for the vertices and edges
    of x.
  SetWrite (wv, we CondOp)

// Returns the write functions for the vertices and edges of x.
  Writes() (CondOp, CondOp)

// x is written on the screen by means of the actual write functions.
  Write()

// Pre: x.Name was called.
// Returns the corresponding file.
  File() pseq.PersistentSequence

// Pre: x.Name was called.
// x is loaded from the corresponding file.
  Load()

// Pre: x.Name was called.
// x is stored in the corresponding file.
  Store()

// Pre: x is connected.
// Returns true, iff x is a ring.
// If x is not connected, true is returned, iff every connected
    component is a ring.
  IsRing() bool

// Pre: v is atomic or imlements Object.
//      e == nil or e is of type uint or implements Valuator;
// Returns an empty graph,
// x is directed, iff d (i.e. otherwise undirected).
// v is the pattern vertex of x defining the vertex type of x.
// For e == nil, e is replaced by uint(1) and all edges of x have the
    value 1.
// Otherwise e is the pattern edge of x defining the edgetype of x.
func New (d bool, v, e any) Graph { return new_(d,v,e) }
```

The representation of the abstract data type Graph and the constructor is quite complicated:

```
package gra
import ("µU/ker"; . "µU/obj"; "µU/kbd"; "µU/pseq")

/* vertex                                                        vertex
```

3.5 Collections of Objects

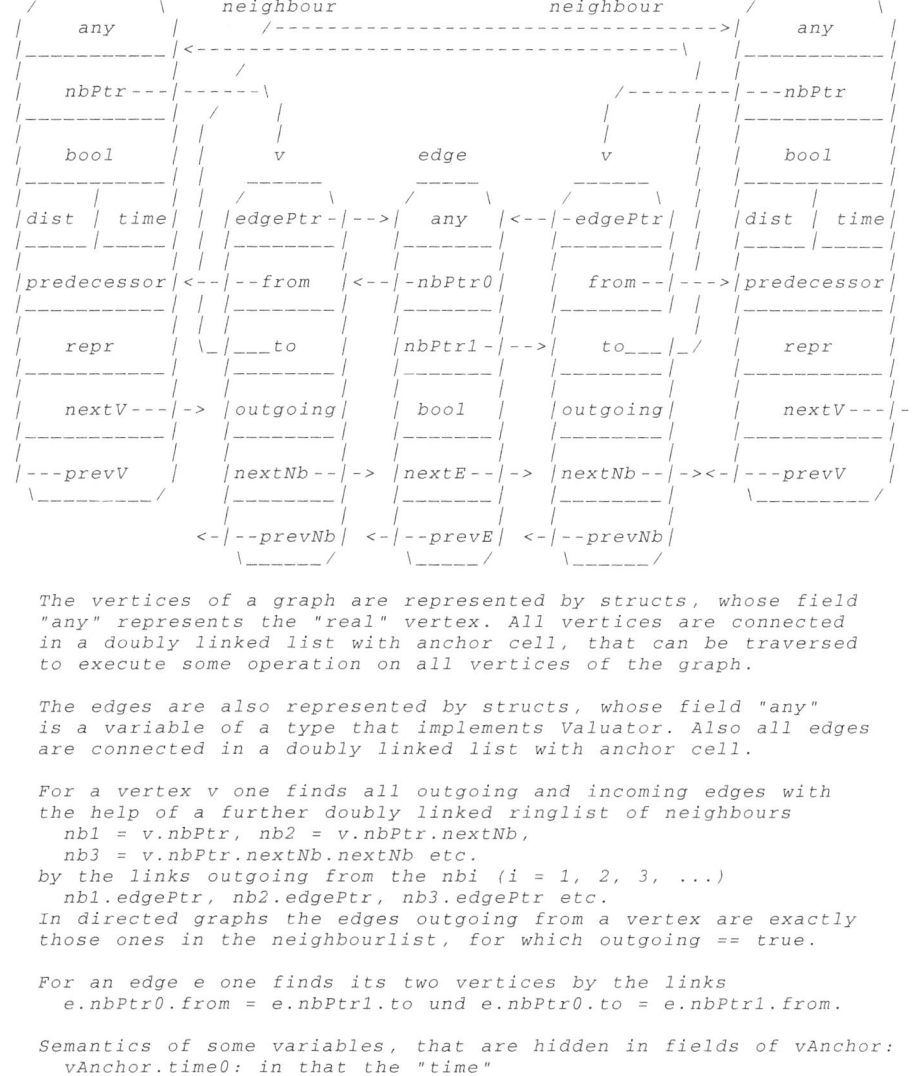

```
The vertices of a graph are represented by structs, whose field
"any" represents the "real" vertex. All vertices are connected
in a doubly linked list with anchor cell, that can be traversed
to execute some operation on all vertices of the graph.

The edges are also represented by structs, whose field "any"
is a variable of a type that implements Valuator. Also all edges
are connected in a doubly linked list with anchor cell.

For a vertex v one finds all outgoing and incoming edges with
the help of a further doubly linked ringlist of neighbours
   nb1 = v.nbPtr, nb2 = v.nbPtr.nextNb,
   nb3 = v.nbPtr.nextNb.nextNb etc.
by the links outgoing from the nbi (i = 1, 2, 3, ...)
   nb1.edgePtr, nb2.edgePtr, nb3.edgePtr etc.
In directed graphs the edges outgoing from a vertex are exactly
those ones in the neighbourlist, for which outgoing == true.

For an edge e one finds its two vertices by the links
   e.nbPtr0.from = e.nbPtr1.to und e.nbPtr0.to = e.nbPtr1.from.

Semantics of some variables, that are hidden in fields of vAnchor:
   vAnchor.time0: in that the "time"
                  is incremented for each search step
   vAnchor.acyclic: (after call of search1) == true
                    iff graph has no cycles. */
const (
  suffix = "gra"
  inf = uint32(1<<32 - 1)
)
type (
  vertex struct {
    any "content of the vertex"
    nbPtr *neighbour
    bool "marked"
    acyclic bool // for the development of design patterns by clients
    dist, // for breadth first search/Dijkstra and use in En/Decode
    time0, time1 uint32 // for applications of depth first search
```

```
    predecessor,//for back pointers in depth first search and in ways
    repr, // for the computation of connected components
    nextV, prevV *vertex
  }

  vCell struct {
    vPtr *vertex
    next *vCell
  }

  edge struct {
    any "attribute of the edge"
    nbPtr0, nbPtr1 *neighbour
    bool "marked"
    nextE, prevE *edge
  }

  neighbour struct {
    edgePtr *edge
    from, to *vertex
    outgoing bool
    nextNb, prevNb *neighbour
  }

  graph struct {
    name,
    filename string
    file pseq.PersistentSequence
    bool "directed"
    nVertices, nEdges uint32
    vAnchor, colocal, local *vertex
    eAnchor *edge
    path []*vertex
    eulerPath []*neighbour
    demo Demoset
    writeV, writeE CondOp
  }
)
type
  nSeq []*neighbour

func newVertex (a any) *vertex {
  v := new(vertex)
  v.any = Clone(a)
  v.time1 = inf // for applications of depth first search
  v.dist = inf
  v.repr = v
  v.nextV, v.prevV = v, v
  return v
}

func newEdge (a any) *edge {
  e := new(edge)
  e.any = Clone(a)
  e.nextE, e.prevE = e, e
  return e
}

func new_(d bool, v, e any) Graph {
  CheckAtomicOrObject(v)
  x := new(graph)
  x.bool = d
  x.vAnchor = newVertex(v)
  if e == nil {
    e = uint(1)
  }
  CheckUintOrValuator (e)
```

```
x.eAnchor = newEdge (e)
x.colocal, x.local = x.vAnchor, x.vAnchor
x.writeV, x.writeE = CondIgnore, CondIgnore
return x
}
```

The suggested algorithms are also very complex; their representation would go far beyond the purpose of this book.

3.6 Additional Data Types from the Microuniverse

Many packages from the microcosm are not presented in this book; these include those

- that implement *number types* (natural, whole, rational, and real numbers), numbers of any precision, *mathematical terms* as well as *vectors* and *matrices*;
- that provide "common types" such as *short strings*, *calendar dates*, *times*, *amounts of money*, and *nations*;
- that deal with two- and three-dimensional *figures* needed for graphics programs—also with the use of OpenGL;
- that deal with *graphs* and their *vertices* and *edges*; and
- that are necessary for the *synchronization of concurrent processes*, which access shared resources.

The last two points are extensively covered in my textbook [1].

References

1. Maurer, C.: Nonsequential and Distributed Programming with Go. Springer Vieweg (2019). https://doi.org/10.1007/978-3-658-26290-7
2. Adelson-Velski, G.M., Landis, J.M.: An algorithm for the organization of information. Sov. Math. **3**, 1259–1263 (1962). English Translation of the Russian Originalarbeit of Ricci, M.J.: http://monet.skku.ac.kr/course_materials/undergraduate/al/lecture/2006/avl.pdf
3. Bresenham, J.E.: Algorithm for computer control of a digital plotter. IBM Syst. J. **4**, 25–30 (1965)
4. Bayer, R., McCreight, E.M.: Organization and maintenance of large ordered indices. In: Proceedings of the 1970 ACM SIGFIDET (Now SIGMOD) Workshop on Data Description, Access and Control, pp. 107–141 (1970). https://doi.org/10.1145/1734663.1734671

Part II
The Projects

General   4

> *Programming is always extending a given system.*
>
> Niklaus Wirth
> From Modula to Oberon,
> Software–Practice and Experience 18 (1988), 661–670

Abstract

This chapter explains what we understand by a *teaching project* and what to pay attention to in them when working in the phases of the software life cycle.

All projects are structured in principle according to the scheme given in the first chapter:

- System analysis.
- System architecture.
- User manual.
- Construction.

The prerequisite for the installation of the teaching projects is the installation of Go and the microuniverse.

You can find instructions for installing Go on the web at `https://maurer-berlin.eu/go`.

You can obtain the microuniverse by downloading the file μU from the page `https://maurer-berlin.eu/mU`, move it with

$$\text{mv } \mu\text{U.tgz go/src}$$

into the subdirectory `go/src` of your home directory, go there with

$$\text{cd go/src,}$$

unpack the microuniverse with

$$\texttt{tar xfzv } \mu\texttt{U.tgz,}$$

and install it with

$$\texttt{go install } \mu.$$

With the call μUou can check if everything worked.

The *source texts* of the projects are in the net under `https://maurer-berlin.eu/obpbook` in the file `o2.tgz` abgelegt. If you have downloaded this file von Ihrem Heimatverzeichnis aus move it with

$$\texttt{mv o2.tgz go/src}$$

into the subdirectory `go/src`, go there with the command

$$\texttt{cd go/src,}$$

and install them with

$$\texttt{tar xfzv o2.tgz.}$$

This will create the subdirectories `rob`, `robi`, `robtest`, `todo`, `life`, `regtest`, `epen`, `mini`, `books`, `inferno`, `lsys`, `bahn`, `rfig`, and `bus`, in which the source texts of the jeweiligen projects are untergebracht.

4.1 Teaching Projects

We understand a *teaching project* to be a small project that is developed for educational purposes in a school or university.

There are *fundamental differences* between the *commercial development* of an IT system and a teaching project. They essentially consist in the fact that

- there is no real "market situation", but rather the character of a teaching and learning situation;
- the participants therefore have more room for design;
- only preliminary experiences in small-scale programming are available;
- the participants work in problematic—fundamentally incompatible—multiple roles, whose tasks in the commercial situation are performed by different people, namely, as
  - learners (with limited and usually not yet stabilized knowledge),
  - system analysts,
  - system architects,
  - designers, and
  - end users or users;

- there is a contradictory tension between
  - necessary complexity for studying typical problems of large-scale programming and
  - sufficient didactic reduction;
- the size of a teaching project is orders of magnitude below that of a commercial project due to limited time resources; and
- an expansion of development capacities (e.g., through overtime or the use of additional employees) is excluded.

The following theses result from this for the management of a teaching project:

- The topic must not be more comprehensive than the detailed examination of a subtask in system analysis.
- The user manual must not degenerate into defining the many interesting ideas that cannot be achieved.
- The specifications must not be less rigid than the participants have learned in small-scale programming.
- The implementations must be possible based on the participants' prior knowledge.
- The project management must limit the task to such an extent that it can guarantee the feasibility of the project, i.e., that
  - they have the "project" running at least as a prototype that includes the essential aspects beforehand,
  - the work can largely rely on existing things, and
  - it is ensured that the participants master these parts to the extent necessary for the work.

It is essential that the participants always have a thorough knowledge of all available partial results: It is repeatedly shown in teaching projects that suddenly individual paths are pursued that do not correspond to the specifications of earlier phases.

4.1.1 System Analysis

The effort for system analysis should usually be kept within narrow limits: The time for deeper elaboration of specialized knowledge is not available and the focus of work in a teaching project should be on informatics issues, even if interdisciplinary or cross-curricular aspects play a role. System analysis in a teaching project can therefore only be understood as a factual analysis of a didactically reduced topic.

To avoid lengthy and unproductive discussions about potential project topics, the topic should be given—if necessary from a well-prepared selection—so that the participants' scope for design is concentrated on the analysis of some object classes and the elaboration of suitable aspects of the task.

In any case, not only the treatment of a new topic, but always also the further development of an existing, well-documented system should be considered, because many of the problems mentioned are solved in a quite natural way.

In this first phase, beginners often underestimate the complexity of the problems to be solved due to their lack of experience, which leads to their expectations of the magnitude of what can be achieved being rarely realistic.

The project management is therefore responsible for estimating the volume of work to be done and thus for calculating the time and personnel resources and their compliance. The planning of unavoidable restrictions is also part of their tasks, as beginners cannot be expected to foresee possible subsequent problems. They must make appropriate considerations in advance of the investigation of suitable topics and ensure that they are taken into account in the task.

Therefore, they must necessarily

- thoroughly investigate topics that are fundamentally suitable for a teaching project for their usability in advance;
- adjust the complexity of the topic to the knowledge to be assumed of the participants;
- ultimately determine the selection of the topic;
- strongly guide the details of the task;
- provide selected parts—fully implemented and documented—to an appropriate extent, so as not to let the participants "reinvent the wheel" every time;
- systematically train the use of these parts; and
- continuously clarify the effects of participants' suggestions during system analysis—if necessary by prototypical work on design and realization.

In addition to the demand for a reasonable manageability of the topic (small topic, even smaller topic, even smaller, much smaller, even smaller), a certain minimum complexity is indispensable to demonstrate typical principles and methods of software engineering and to convey insight into their necessity:

- inclusion of a structure of the involved objects nested over at least three levels, which is branched at at least one point, to achieve a non-trivial depth and branching in the system architecture,
- exemplary creation of reusable components that can also be used for other purposes, as well as
- consideration of the possibility of alternative implementations of certain components.

4.1.2 System Architecture

There are no specific peculiarities of teaching projects in this phase that go beyond what is postulated in Sect. 1.1.2.

The question of how to start the *actual programming activity*, which is very difficult for beginners according to experience, is solved in this phase, because it shows in the work how *astonishingly simple* the system architecture of the system results from a system analysis, the strictly oriented *towards the objects of a system*.

4.1.3 User Manual

The considerations of how the planned system should present itself to the users, especially questions of ergonomic operation, are tedious and time-consuming; they are often controversially discussed and their necessity is not always recognized at the beginning of the work.

It is mandatory for each component used to specify the semantics of the data objects and the complete and contradiction-free specification of all access operations they are often controversially discussed, stating all prerequisites and effects. As a rule, clean colloquial formulations are sufficient for this.

The implementations of the representation of the data types and the access operations in the components are based on the knowledge acquired in small-scale programming, which may need to be supplemented or expanded by studying the relevant literature. This also serves to secure and exemplify the deepening of the corresponding skills and abilities.

If design errors become apparent during the implementation of a component (usually in the form of incompleteness or lack of clarity in its specification), the specification is corrected in agreement with all involved clients—in any case only after consultation with the project management—and its implementation is adapted to the changes. A prerequisite for system integration is of course the systematic test of the developed components as part of the construction.

Robi 5

> *keep going:*
> *If not yet at the edge of the world, then*
> *move one place further and*
> *keep going*

Abstract

The program presented here was used in the teacher training course in computer science at the Free University of Berlin at the beginning of the lecture on imperative and object-oriented programming to make the transition from the functional to the imperative paradigm as smooth as possible—through the basic idea of *variable-free programming*.

Robilanguage is a very simple language that is suitable for developing the basic concepts of imperative programming.

The significance of this concept lies in the

- *variable-free introduction to structured programming*:
  - without any baroque feature of any imperative language (i.e., in principle without a definitive commitment to a specific language paradigm);
  - therefore easy to learn, but still immediately at the heart of computer science, i.e., dealing with typical computer science problems;
  - with a wealth of possible exercises that can also be solved with "pencil and paper";
  - the use of computers is limited to the "experiment" to confirm or refute theses;
- therefore, working on the computer consists only of short periods of time.
- *specification of methods/functions*:
  - precise specification (in the form of static state descriptions) of the preconditions for their call and the effects after their execution.

- *recursion*:
  - as a central language tool of every programming language (besides sequence and case distinction);
  - before the introduction of *iteration*, i.e., dealing with the different types of *loops* (pre-testing, post-testing, counting).
- Introduction of *parameters* in methods/functions:
- Concept of *components*:
  - for using services only through their specification according to the *principle of secrecy*, i.e., without disclosing their implementation;
  - here using the example of the programming language Go, which is excellently suited for this due to the possibility of *strict textual separation* of specification and implementation.
- *depth-first search*:
  - in connection with algorithms for *backtracking*.

The problem is to have Robi move around in this block world according to certain rules, and to have him lay down, pick up, move, or count blocks according to certain principles, or to block squares by walling them up.

Since there are no limits to the imagination for inventing exercises that Robi should solve, and for rules that he must apply, there is an inexhaustible field of exercises, which is excellently suited both as motivation for getting started in various aspects of computer science and for introducing demanding concepts.

5.1 System Analysis

Robi is a "robot" that "lives" in a rectangular world made of checkerboard-arranged squares. He always stands on one of the squares and looks in one of the four *cardinal directions*. Robi can turn *90 degrees* to the *left* or *right* and take a *step* (i.e., move one square) in the direction he is looking (unless he is already at the edge of his world). Therefore, in principle, every square in the world is accessible to him.

On each square, one or more *blocks* can be placed. Robi carries a bag of blocks with him. He can *place* a block from his bag onto the square where he stands, put (as long as he still has blocks) or *take* a block from any square where he stands, remove it, and put it in his bag (as long as there is still one there). He is therefore able to occupy any squares in the world (within the scope of his available blocks) with blocks.

Robi can *mark* squares and also *remove* the markings, so he is able to remember where he has already been. This allows him to search specifically (*depth search*!). Robi can also *wall up* squares so that they are no longer accessible, and *remove* the walls again.

In addition, Robi can also *push* blocks (if the square behind is empty). Therefore, you can also let Robi play *sokoban*, provided, you have thought up appropriate worlds and created them with the *robi editor* (see below).

The problem is to have Robi move around in this block world according to certain rules, and to have him lay down, pick up, move, or count blocks according to certain principles, or to block squares by walling them up.

Since there are no limits to the imagination for inventing exercises that Robi should solve, and for rules that he must apply, there is an inexhaustible field of exercises, which is excellently suited both as motivation for getting started in various aspects of computer science and for introducing demanding concepts.

5.2 The Robi Language

The robi language consists of

- statements
  - TurnLeft
  - TurnRight
  - Run
  - RunBack
  - PickUp
  - PutDown
  - Push
  - Mark
  - Unmark
  - WallUp
  - WallDown
- predicates
  - InUpperLeftCorner
  - AtEdge
  - Empty
  - NeighbourEmpty
  - HasBlocks
  - Marked
  - NeighbourMarked
  - InFrontOfWall
- and a counting function
  - NumberOfBlocks

as well as composite instructions

- Sequences ... ; ...
- case distinctions `if` ... `{...}` `else` `{...}`

5.3 System Architecture

It is very flat, there is only *one* abstract data type, the `robot`, and the abstract data object in the package `robi`.

5.4 User Manual

There are the three programs mentioned above, which will be introduced in the following.

5.4.1 The Robi Editor

With the program `robiedit`, a new robot world can be created and an existing one can be modified. The name of the world can be given as a parameter to the program call; without a parameter, it is named "World" (the world files have the suffix `.rob`). Robi's *place* and his *direction* are evident from the direction of the figure. Robi performs the following actions when a *command key* is pressed:

- Arrow key ▲, ▼, ◀ or ▶: If this is the key in the direction of his view, he *runs* one space further, otherwise he changes the *direction of his view* in the direction of the arrow. Beforehand, it must be checked whether he is standing at the edge of the world and whether the space in front of him is walled up.
- Insert key Ins: If Robi still has a block in his bag, he places a block in his space.
- Del: If there is a block in Robi's space, he picks it up and puts it in his bag.
- Enter key ↵: If there is a block in the space in front of Robi and the space behind it is free, he pushes the block into that space and stands there.
- Backspace key ←: The last action is undone.
- Pos1: If the space in front of Robi is free, he walls up his space and stands on it.
- End: If there is a wall in the space in front of Robi, he tears it down and stands on that space.
- F5: Robi marks his space.
- ⇆: If the space where the mouse pointer is pointing is free, a robot is placed there; if there is a robot there, it is removed from the world.
- F6: If Robi's space is marked, the mark is removed.

5.4 User Manual

- F1: The key assignment for control is displayed.
- Esc: The program is terminated. (Robi's world is available in its current state at the next program run under the same name.)

5.4.2 The Robi Protocol

The program `robiprog`:

- Operation like `robiedit`, furthermore
- a *protocol* is generated in the form of a program file, i.e., all editing steps are logged in the form of a source code for a Go program. It is important to ensure that the generated program is started with the same state of the world with which `Robiprog` was started (because the world was changed with Robiprog).

5.4.3 Robi-Sokoban

The program `robisoko`:

- Operation: greatly simplified version of `Robiedit`, suitable for playing Sokoban (= pushing all blocks onto the (same number of) marked places in a world like, for example, *Sokoban1*).

5.4.4 Robot Race

In the program `robrace`, as many robots race two "rounds" as were given as parameters to the program call (at least 2, at most 24). The ⏎ key must be held down until the race is over.

5.4.5 General Procedure

- A *Robi program*, i.e., a program using the package `robi`, is *written*.
- It is *translated* and *linked* with the call "`go install`".
- Step by step *executed* with ⏎.
- *Terminated* with Esc.
- In case of problems, *aborted* with the combination Ctrl C.

5.5 Construction

Here is the specification of the robot package `rob`:

```
package rob
/* Manages robots that live in a rectangular world of places arranged
   in a checkerboard pattern. The world is 24 places wide and high.
   Blocks can lie on the places or the places can be walled up.

   Every robot stands always on one of these places that is designated
   as "R's place". It always stands in one of the four celestial directions
   that are designated as R's direction.

   Every robot has a pocket with initially Max blocks and
   always has access to bricks.

   The places of the robots, their directions and the number of blocks
   in their pockets are the same as the last time the wordl was called.
   If the world is new, it is empty and a robot stands in direction south
   in the northwest corner and has Max blocks in his pocket,
   minus those that he already has placed in the world.

   Initially, the protocol is not switched on.

   The calling robot is always designated as "R". */

import . "µU/obj"

const (M = 24    // number of places per row and column of the world
       Max = 999) // maximal number of blocks in R's pocket

type Robot interface {

  Coder

// Returns the number of R.
  Number() uint

// R has turned left 90$^\circ$ to the left.
  TurnLeft()

// R has turned left 90$^\circ$ to the right.
  TurnRight()

// Returns true, iff R stands in the northwest corner.
  InUpperLeftCorner() bool

// Returns true, iff R's place has no neighbourplace in R's direction.
  AtEdge() bool

// Pre: R does not stand at the edge and
//      the neighbourplace in R's direction is not walled up.
// R stands in the same direction as before on this neighbourplace.
  Run()

// Pre: R's place has entgegen R's direction a neighbourplace,
// that is not walled up.
// R stands in the same direction as before on this neighbourplace.
  RunBack()

// Returns true, if there are no blocks on R's place.
  Empty() bool

// Returns true, if R's place in R's direction has a neighbourplace
// and on this place does not lie a block.
  NeighbourEmpty() bool
```

5.5 Construction

```
// Returns true, if R's pocket is not empty.
  HasBlocks() bool

// Returns the number of blocks in R's pocket.
  NumberOfBlocks() uint

// Pre: R's Tasche is not empty.
// On R's place lies on block more than before,
// in his pocket is one less.
  PutDown()

// Pre: On R's place lies at least one block.
// On R's place lies one block less than before,
// in his pocket is one more.
  PickUp()

// Returns true, iff R's place in R's direction has a neighbourplace,
// on which lies exactly one block and this neighbourplatz in turn has
// a neighbourplace in R's direction that is empty and not walled up.
// In this case R stands in the sacme direction as before on the
// previous neighbourplace in its direction and the block that was
// not on it before is not on the neighbourplace in R's direction.
  Pushed() bool

// Pre: R's place has in R's direction a neighbourplace,
//      on which exactly one block lies.
// This block is as far as possible (i.e., without collision
// with robots, blocks or walls) in R's direction weitergeschoben.
  Push()

// R's place is marked.
  Mark()

// R's place is not marked.
  Unmark()

// Returns true, iff R's place is marked.
  Marked() bool

// Returns true, iff R's place in R's direction
// has a neighbourplace that is marked.
  NeighbourMarked() bool

// Returns true, if R's place in R's direction
// has a neighbourplace that is walled up.
  InFrontOfWall() bool

// Pre: R's place has in R's direction a neighbourplace,
//       that is not walled up.
// R stands in the same direction as before on this neighbourplace
// If on the place on which R stood before were blocks, now they
// do not lie there any more, but are in his pocket; but this place
// is now walled up. A mark previously present there mark is now removed.
  WallUp()

// Pre: R's place has in R's direction a neighbourplace,
//       that is walled up.
// R stands in the same direction as before on this neighbourplace
// and this place now is not walled up.
  WallDown()

// The robot world is written to the screen.
  Write()

// Returns the position of R.
  Pos() (uint, uint)
```

```
  // R's position is (x, y).
  Set (x, y uint)
}

// The robot world is that one whose name was given
// at the call of the programm as parameter.
func Load (s ...string) { load(s...) }

func NumberRobots() uint { return uint(nRobots) }

// The blocks and bricks in the robot world now lie
// on the places specified by the user.
// If R is standing on a place where there are blocks,
// their number is shown. R's place and direction,
// the number of blocks in its pocket and of the blocks
// on all places are at the next program run with this world
// the same as for the call of this method.
// If the protocol is switched on, the editing process
// is protocolled in a Go-source text (under the name
// of the robot world with the suffix ".go").
// The program generated from this source text by translation
// simulates step by step the editing process
func Edit() { edit() }

// Pre: x < M, y < M.
// Returns a new robot on the position (x, y)
// with Max blocks in its pocket.
func NewRobot (x, y uint) Robot { return newRobot(x,y) }

// Pre: n > 0. There is at least one robot in the world.
// Returns the robot with the number n.
func Nr (n uint) Robot { return all[n] }

// The robot world is written to the scren.
func WriteWorld() { writeWorld() }

// After editing the program is generated
// that reproduces the editing process.
func GenerateProgram() { generateProgram() }

// The protocol is switched on, iff b == true (see edit).
func SwitchProtocol (b bool) { switchProtocol(b) }

// For b == true the behaviour of the editor is simplified
// according to the requirements of the game Sokoban.
func SwitchSokoban (b bool) { switchSokoban(b) }

// n is written to the screen in the bottom line of the screen
// The calling process was then stopped until the user
// has acknowledged the output with Esc.
func Output (n uint) { output(n) }

// Returns the number entered by the user
// in the bottom line of the screen.
func Input() uint { return input() }

// s and n are written in a line at the bottom of the screen.
// The calling process was then stopped
// until the output was acknowlegded with Esc.
// Now the error report is removed from the screen.
func ReportError (s string, n uint) { reportError(s,n) }

// s and n are written in a line at the bottom of the screen.
func Hint (s string, n uint) { hint(s,n) }
```

```
// The program has ended with the error report ("program has ended")
func Ready() { ready() }
```

The "Robi" is the abstract data object robi; it consists of an instance of an object of type Robot; and its specification is thus quasi-identical with a part of the specification of the package `rob`:

```
package robi
// Specifications% see µU/rob/def.go
func M() uint { return m() }
func TurnLeft() { turnLeft() }
func TurnRight() { turnRight() }
func InUpperLeftCorner() bool { return inUpperLeftCorner() }
func AtEdge() bool { return atEdge() }
func Run() { run() }
func RunBack() { runBack() }
func Empty() bool { return () }
func NeighbourEmpty() bool { return neighbourEmpty() }
func NumberOfBlocks() uint { return numberOfBlocks() }
func HasBlocks() bool { return () }
func PutDown() { putDown() }
func PickUp() { pickUp() }
func Pushed() bool { return pushed() }
func Push() { push() }
func Mark() { mark() }
func Unmark() { unmark() }
func Marked() bool { return marked() }
func NeighbourMarked() bool { return neighbourMarked() }
func InFrontOfWall() bool { return inFrontOfWall() }
func WallUp() { wallUp() }
func WallDown() { wallDown() }
func Load (s ...string) { load(s...) }
func Edit() { edit() }
func SwitchProtocol (b bool) { switchProtocol(b) }
func SwitchSokoban (b bool) { switchSokoban(b) }
func Write (n uint) { write(n) }
func Input() uint { return input() }
func ReportError (s string, n uint) { (s,n) }
func Hint (s string, n uint) { hint(s,n) }
func Ready() { ready() }
func Pos() (uint, uint) { return pos() }
func Set (x, y uint) { set(x,y) }
```

5.6 Exercises

Sought are variable-free and recursive implementations

- of the following problem of the "pliant guard" by Nievergelt (see [1]):
  The world houses a medieval city. This is protected by an arbitrarily complex city wall. Robi starts within the city wall and is to patrol along the wall forever as a reliable guard, in such a way that he could always touch the wall with his right hand, if he had one. The city is available as Robiworld under the name `city`.
- of the following more challenging set of problems: Robi is to be in a given world (also with walls)

- find a block in a *maze* (the maze is available under the name `maze`),
- find *all* blocks in a maze and count them.
  Note: This is the algorithm of *depth-first search* by backtracking; for this, it is necessary to mark all already visited places.

After introducing the concept of variables and iteration through loops, for all exercises

- the *iterative* versions of all *recursive* function calls

are sought. You can compare your solutions with the sample solutions given in the next section.

Figures 5.1 and 5.2 show the city from the first exercise and the maze from the second exercise.

5.6.1 Sample Solutions

First exercise

```
package main
import . "robi"

func toTheWall() {
  if ! InFrontOfWall() {
    Run()
    toTheWall()
  }
}

func guard() {
  if InFrontOfWall() {
    TurnLeft()
  } else {
    Run()
    TurnRight()
    if ! InFrontOfWall() {
      Run()
    }
  }
  guard()
}

func main () {
  Load ("city")
  toTheWall()
  guard()
}
```

Second Exercise

```
package main // depth search
import . "robi"

var numberOfBlocks uint

func ok() bool {
  return ! AtEdge() && ! InFrontOfWall()
```

5.6 Exercises

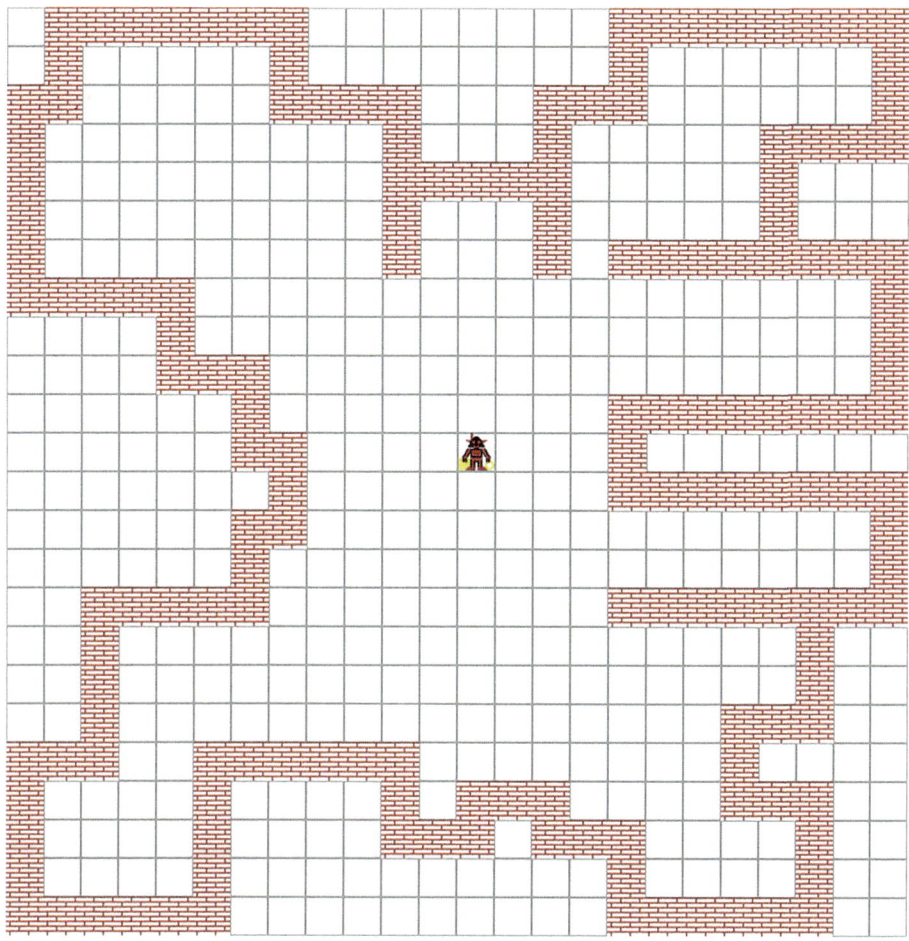

Fig. 5.1 The city from the first exercise

```
}
func search() {
  var leftOk, straightOk, rightOk bool
  if Marked() {
    return
  }
  Mark()
  if ! Empty() {
    numberOfBlocks++
    Hint ("number of blocks =", numberOfBlocks)
  }
// find out, in which directions Robi can continue run:
  TurnLeft()
  leftOk = ok()
  TurnRight()
  straightOk = ok()
```

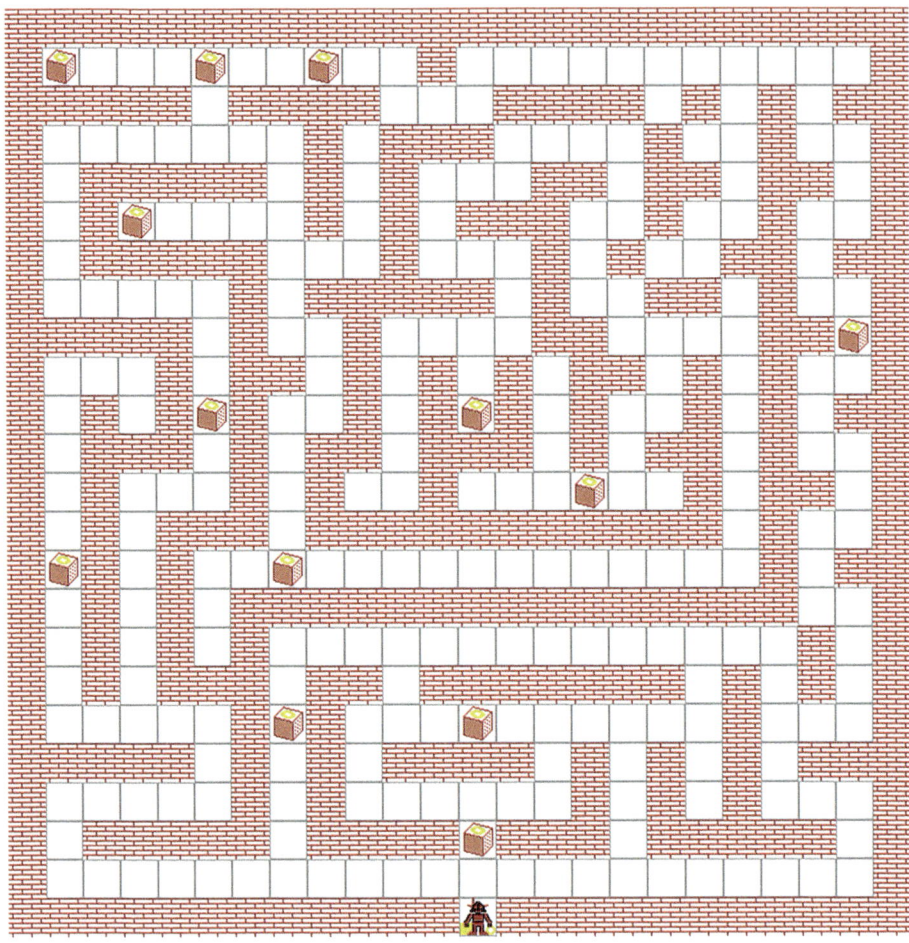

Fig. 5.2 The maze from the second exercise with 13 blocks

```
  TurnRight()
  rightOk = ok()
  TurnLeft()
// start of the depth search
  x, y := Pos()
  if leftOk {
    TurnLeft()
    Run()
    search()
    TurnRight()
  }
  Set (x, y)
  if straightOk {
    Run()
    search()
  }
  Set (x, y)
```

```
  if rightOk{
    TurnRight()
    Run()
    search()
    TurnLeft()
  }
}
func main() {
  Load ("maze")
  search()
  ReportError ("number of blocks =", numberOfBlocks)
  Ready()
}
```

References

1. Nievergelt, J.: Roboter programmieren - ein Kinderspiel. Informatik-Spektrum, Heft **22**, 364–375 (1999)

The Appointment Calendar 6

Yes, just make a plan,
be just a great light!
And then make a second plan,
both of them won't work.

Berthold Brecht
From The Threepenny Opera

Abstract

This educational project focused on creating a personal scheduler with the aim of being able to enter, modify, delete, reschedule, and copy appointments. With the help of a suitable keyword system, it should also be possible to specifically search for all appointments in which these keywords appear.

Two consecutive courses of teacher training in computer science at the Free University of Berlin dealt with this topic. The task was to construct an "electronic" appointment calendar for managing the appointments of a single person and to be able to find appointments with certain keywords in it. Particular emphasis was placed on ergonomic design—a clear presentation and easy operability.

6.1 System Analysis

In this section, we present the results of the investigations on which objects appear in the appointment calendar.

6.1.1 Calendar Pages

For reasons of clarity, all *appointments of a day* should be ordered chronologically and be comprehensible at a glance, i.e., they must be accommodated on one screen page. This—day-oriented—summary of appointments is referred to as a *calendar page*; the *appointment calendar* is then a *sequence of these calendar pages*. The *actual* calendar page is always the one that is currently visible on the screen.

A *calendar page* consists of

- a *calendar date*,
- a *day attribute*, and
- the *sequence of appointments* of the respective day.

A calendar page is accessed via its date. In addition to the date, the respective day of the week and possibly the day attribute are output.

The entry, modification, and deletion of appointments take place within the calendar pages. New appointments are always entered in the first free line of a calendar page. To avoid a rather confusing unrest on the screen, newly entered appointments are only sorted chronologically when the same calendar page is called up again.

6.1.2 Day Attributes

The *day attributes* serve the purpose of being able to quickly overlook certain days or periods in the appointment calendar. They should be indicated by a short text or the visual highlighting of the corresponding calendar dates.

You can configure any day attributes by creating the text file `dayattributes.cfg` in the subdirectory `.todo` of your home directory, in which you line by line deposit the words for the day attributes that you need. The first line *must* be "`keyword`". Example:

```
keyword
vacation
```

The name `dayattributes.cfg` is mandatory, unless you change the source code of the program.

6.1.3 Sequences of Appointments and Appointments

The appointments of a calendar page should be listed in *chronological order*, including simultaneous appointments. Appointments *without* a time indication should be at the *beginning* of this order, as they are non-time-bound references to special events such as birthdays.

6.1 System Analysis

The other appointments are sorted by time; if there are multiple appointments at the same time, they are sorted by appointment attributes, possibly by keywords or texts.

An appointment consists of

- a *date*,
- a *time*,
- an *appointment attribute*,
- a *keyword*, and
- a *text*.

The *date* of an appointment is given by the day of the calendar page in which it appears.

Appointments can be classified by *appointment attributes* to enable users to set priorities in planning and attending their appointments and to facilitate the management of appointments of a certain category and the search for them.

The appointment attributes in the appointments of a calendar page are represented by *[abbreviations* of *three* characters, preferably with different initial letters. You can configure any appointment attributes by creating the text file `appointmentattributes.cfg` in the subdirectory `.todo` of your home directory, in which you list the abbreviations for the appointment attributes you need, line by line. Example:

```
wrk
bid
prv
```

for *work*, for *birthday*, and for *private*. The name appointmentattributes.cfg is mandatory.

The length of the *text* of an appointment is designed to accommodate as many characters as will fit on one line of the screen with all components of the appointment.

Since in a *unstructured* text, keywords cannot be easily identified as such, the *text* is divided into a short *keyword* as a *search term* and the other—not further structured—information to facilitate the search for specific appointments. The search is also conducted for parts of the keyword and regardless of case, to find, for example, "Conf" as well as "Soc.-Conf." or "Subject conference".

For the "transport" of appointments between different calendar pages, a storage—an invisible buffer area—is provided.

Appointments of a calendar page can be *copied into the storage* (and also *deleted* from the calendar page) or *inserted* from the storage into the current calendar page, as long as there is still space (while the content of the storage remains).

This allows appointments to be moved to any day or copied to multiple calendar pages, e.g., for easy entry of recurring appointments at the same times.

6.1.4 Annual Calendar

For a quick overview, an *annual calendar* can be displayed for each date, grouped by months, with the day numbers in columns side by side each week in each month. Such an annual calendar can be displayed in a rectangle of 25 lines and 80 columns, which determines the size of the screen.

In it, the *Sundays and holidays* are *easily recognizable*. Figure 6.1 shows it.

The annual calendar is also used to *call up the calendar pages*: *exactly one day* in it must be marked in a clear way as the *actual day*. This marking, i.e., the current day, can be moved at will.

A control system is provided for selecting a date, which is determined by the temporal structure of the year: It is possible to jump between calendar dates by day, week, month, quarter, and year.

6.1.5 Monthly and Weekly Calendars

For a rough overview of the *appointments of a month*, a *monthly calendar* should be able to be output, in which the days on which appointments are entered are displayed with references to all appointment attributes occurring on them.

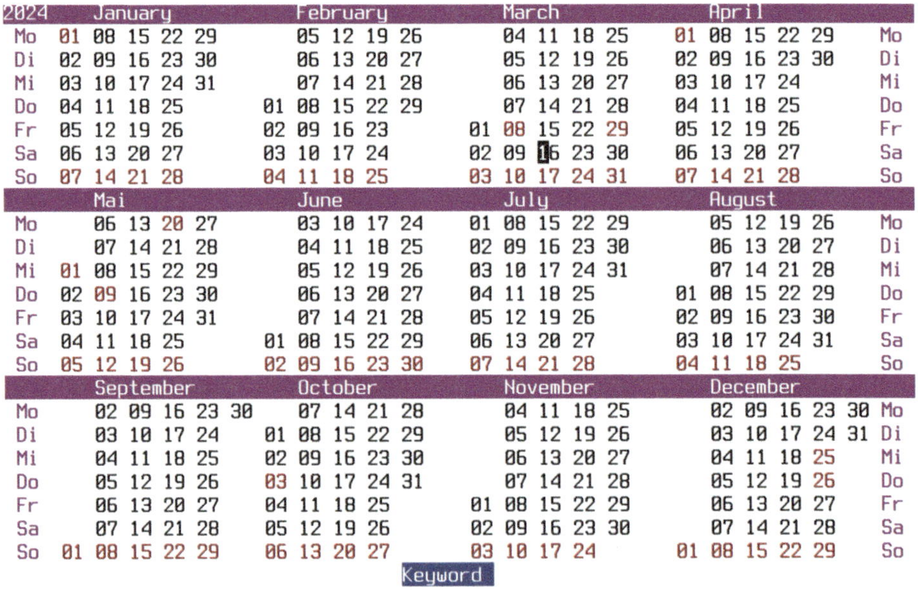

```
2024      January            February           March              April
Mo    01 08 15 22 29      05 12 19 26        04 11 18 25      01 08 15 22 29      Mo
Di    02 09 16 23 30      06 13 20 27        05 12 19 26      02 09 16 23 30      Di
Mi    03 10 17 24 31      07 14 21 28        06 13 20 27      03 10 17 24         Mi
Do    04 11 18 25      01 08 15 22 29        07 14 21 28      04 11 18 25         Do
Fr    05 12 19 26      02 09 16 23        01 08 15 22 29      05 12 19 26         Fr
Sa    06 13 20 27      03 10 17 24        02 09 16 23 30      06 13 20 27         Sa
So    07 14 21 28      04 11 18 25        03 10 17 24 31      07 14 21 28         So
          Mai                June               July               August
Mo       06 13 20 27      03 10 17 24      01 08 15 22 29      05 12 19 26         Mo
Di       07 14 21 28      04 11 18 25      02 09 16 23 30      06 13 20 27         Di
Mi    01 08 15 22 29      05 12 19 26      03 10 17 24 31      07 14 21 28         Mi
Do    02 09 16 23 30      06 13 20 27      04 11 18 25      01 08 15 22 29         Do
Fr    03 10 17 24 31      07 14 21 28      05 12 19 26      02 09 16 23 30         Fr
Sa    04 11 18 25      01 08 15 22 29      06 13 20 27      03 10 17 24 31         Sa
So    05 12 19 26      02 09 16 23 30      07 14 21 28      04 11 18 25            So
        September            October           November           December
Mo    02 09 16 23 30      07 14 21 28      04 11 18 25      02 09 16 23 30 Mo
Di    03 10 17 24      01 08 15 22 29      05 12 19 26      03 10 17 24 31 Di
Mi    04 11 18 25      02 09 16 23 30      06 13 20 27      04 11 18 25         Mi
Do    05 12 19 26      03 10 17 24 31      07 14 21 28      05 12 19 26         Do
Fr    06 13 20 27      04 11 18 25      01 08 15 22 29      06 13 20 27         Fr
Sa    07 14 21 28      05 12 19 26      02 09 16 23 30      07 14 21 28         Sa
So 01 08 15 22 29      06 13 20 27      03 10 17 24      01 08 15 22 29         So
                                    Keyword
```

Fig. 6.1 The annual calendar with entered vacation times

- the *Sundays and holidays*,
- the days with *set attributes*, and
- the days on which *sought* appointments were found.

For a somewhat more precise overview of the *appointments of a week*, a *weekly calendar* is also provided. With a division of the screen into seven columns—which is obvious in view of the appearance of the calendar pages—there is room in it for each appointment's time and a (sufficiently short) representation of its *appointment attribute*.

The actual day can be set in these calendars as in the annual calendar.

In the business world, appointments tied to weeks are defined by the *week number*, which is also output in these overviews (according to DIN 8601, the 1st week is the one in which the first Thursday of the year falls).

In them,

- the *Sundays and holidays*,
- the days with *set attributes*, and
- the days an denen *gesuchte appointments*

should be *easily* recognizable through optical highlights, which clearly differ from each other for each of these groups. This makes the particularly important days or periods for the users immediately readable at a glance.

However—apart from the holiday attribute—only *one* day attribute is displayed at a time, so as not to confuse by overlaying too many pieces of information. It should be possible to switch cyclically between the configured day attributes in order to overlook the various attributed periods one after the other.

The *actual day attribute* can be *set* and *deleted* for each day, because this is considerably more practical for periods than doing it day by day in the individual calendar pages.

6.1.6 Appointment Calendar

For obvious ergonomic reasons, the representations of the different calendars on the screen (e.g., days in columns by week, appointments one below the other) are largely adapted to each other, i.e., the optical highlighting of Sundays and holidays and the day attributes should be the same in all cases and the basic operation of the system when switching between annual, monthly, and weekly calendars and when manipulating the day attributes should be uniform.

After calling up the program, the annual calendar should be shown; the actual day is the system date of the used computer. From there the monthly calendar, then the weekly calendar, and finally a calendar page are switched to and vice versa.

The actual day remains actual until it is not changed, i.e., for example, it is possible to *flip* forward and backward through the appointment calendar at the level of the calendar pages. This also facilitates the transfer of appointment entries, for which the direct switching option between different days is needed.

6.1.7 Search for Appointments

From the monthly or weekly calendar, a search term can be entered. If this search term is part of a search word in an appointment, the search word is conspicuously marked on all calendar pages with these appointments and in the weekly and monthly calendars, conspicuous marks are set on the days on which such an appointment exists.

6.2 System Architecture

6.2.1 The Objects of the System

Each of these objects forms a data type, which is "packaged" in a package. Thus, we have the packages

- `day` for calendar data,
- `clk` for times,
- `todo/attr` for the appointment attributes,
- `todo/word` for the keywords,
- `text` for the texts,
- `todo/appt` for the appointments,
- `todo/appts` for the appointment sequences,
- `todo/dayattr` for the day attributes,
- `todo/pdays` for persistent sets of calendar data,
- `todo/page` for the calendar pages, and
- `todo/cal` for the sequence of calendar pages.

6.2.2 Component Hierarchy

These packages depend on each other as shown in Fig. 6.2, where the lower package is used (imported) by the one above. The packages day, clk, and text are components of the microuniverse.

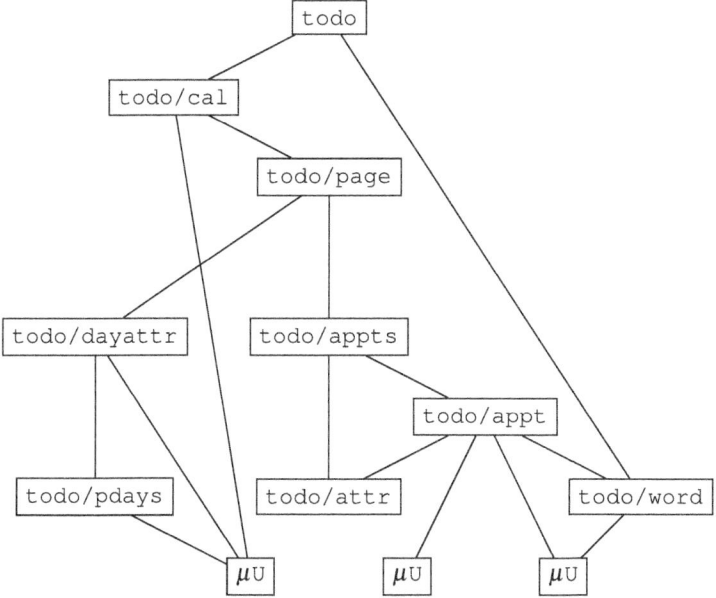

Fig. 6.2 System architecture of the appointment calendar

The packages day, clk, and text are components of their microuniverse. At lower levels, many other packages are needed, which due to their universal usability are also components of the microuniverse and which we introduced in Chap. 3:

- box for input/output fields,
- seq for sequences,
- pseq for persistent sequences,
- set for sets, and
- piset for persistent sets.

6.3 User Manual

After the categorization in the third paragraph of Sect. 1.1.3 about the user manual, it consists of two parts:

- the specification of the *formats* of the objects that appear on the screen and
- the *system operation*.

6.3.1 Formats

In this first part, the formats of all objects are defined. The corresponding specifications are completely independent of the control of the system, which is described in Sect. 6.6.2.

For calendar dates, the formats Dd, Dd_mm_, Dd_mm_yy, Yyyy, Wd, WD, M, and WN are used (see specification in μU).

Times consist of hours and minutes, separated by a dot; the "timeless" time is represented only by spaces.

6.3.1.1 Appointment Attributes

Appointment attributes have two formats:

- the long format consisting of three characters or
- the short format, only the initial letter of the long format.

The reason for this decision lies in the formatting of the weekly calendar (see Sect. 6.5).

The use of the short format requires that the appointment attributes all must have different initial letters.

If you have configured the appointment attributes as in Sect. 6.1.3, their initial letter is sufficient for input (see above). Spaces mean renunciation of the specification of an appointment attribute.

6.3.1.2 Keywords

A *keyword* is provided with 12 characters.

6.3.1.3 Texts

The *text* of an appointment can be 56 characters long, because out of the 80 characters of a screen line, 5 are used for the time, 3 for the appointment attribute, 12 for the keyword, and 4 spaces for separation and ending.

6.3.1.4 Appointments

From the previous considerations, it follows that all attributes of an appointment, separated from each other by a space, fit into a screen line with 80 characters. This allows a good 20 appointments to be accommodated one below the other on a screen with 25 lines, which optimally guarantees the demand for clarity.

For the representation of an appointment in the weekly calendar, 9 characters remain (on a screen with 80 columns): (7 days per week side by side, 2 spaces for horizontal separation).

Therefore, two formats are provided for appointments:

- the *long format* with all attributes, each separated by a character, with a width of 79 characters and
- the *short format* only with time and appointment attribute directly next to each other (distinguished by colour) in a width of 9 characters.

6.3.1.5 Day Attributes

Day attributes are displayed in two formats:

- as a word with up to eight letters or
- by *colour highlighting* the relevant calendar dates with a different background colour.

In the annual, monthly calendar, and weekly calendar, the word of the *actual* day attribute is output in the last screen line and the calendar dates of the relevant days are colour-marked. When searching for appointments (see Sect. 6.6.1), the day attribute "keyword" is actual.

6.4 Calendar Pages

The structuring of the calendar page on the screen results from the previous considerations:

In the first line on the left edge is the weekday of the *actual date*, behind it the actual date in the format Dd_mm_yyyy, in the middle the *day attribute* of the relevant day and below it, separated by a blank line, in the 3rd to 23rd line, the *sequence of appointments*, consisting of 21 *appointments* in long format.

The last screen line remains free for instructions for use or error messages.

6.5 Weekly Calendar

The *weekly overview* consists of the appointments of a week in *short format* (see Sect. 6.3.1.4), column-wise for all days of the week side by side.

The appointments of each day are listed one below the other; each column is headed with the date of the respective day in the format Wd Dd_mm_.

In the first row, the year is displayed on the left and right outer sides, and the week number (date in the format WN) is in the middle.

In the middle of the last screen line, the word of the *actual day attribute* is displayed; the calendar data of the respective days are highlighted in colour.

6.5.1 Monthly Calendar

The *monthly overview* consists of the calendar data of the month in the format Dd_mm_, column-wise from Monday to Sunday; on the right and left at both screen edges—matching in the respective row—the weekdays are in the format Wd.

In the row below, there is a string that consists of the sequence of the *appointment* attributes occurring on this day in the *short format* (see Sect. 6.7.2).

In the first row, the year is displayed on the left and right outer sides, and the month name (date in the format M) is in the middle; the *actual day attribute* is displayed as in the weekly calendar.

6.6 Annual Calendar

The *annual calendar* fits on a screen with 25 lines and 80 columns:

When displaying the weeks column-wise in the format Dd, 8 lines (month name and 7 d per week) are needed for a month block with a maximum of 6 weeks (e.g., first of the month on Saturday, last on Monday) 6 times 3 columns are needed (2 digits for the day and a space). This results in a display in the form of 3 times 4 month blocks side by side, for which 24 lines and 72 columns are needed.

In the upper left corner of the screen, the year is displayed; on the left and right screen edges, the abbreviations of the weekdays are displayed in the format Wd. The current day attribute is displayed as in the weekly calendar.

6.6.1 Search and Search Results

The *search term*, which the appointment calendar is searched for, has the same format as the keywords. The corresponding field is located in the weekly, monthly, and annual calendar next to the word of the current day attribute in the last screen line.

After entering a search term in the weekly, monthly, or annual calendars, the search results appear, i.e., those days on which appointments are entered, in whose keyword the search term is contained, are highlighted in the calendars in the same colour as the days of certain day attributes; in addition, the search words in the corresponding appointments are highlighted in colour (this highlighting remains when switching to the daily calendar).

Consequently, during the search, the current day attribute is reset to "search word" so that the highlights are clear.

6.6.2 System Operation

This second part describes how the appointment calendar is controlled by users.

At the start of the program, the annual calendar is shown. The screens can be cycled through in the order

1) Year,
2) Annual calendar,
3) Monthly calendar,
4) Weekly calendar, and
5) Calendar page

you can move one step forward with ◄┘ and one step back with Esc—with the exceptions that at 1) the program is exited with Esc and that at 5) it does not continue with ◄┘. In cases 1) to 4) the cursor blinks in the field for the actual date; it can be changed, after which the screen is updated to the year, month, week, or day that was entered.

At the beginning, no day attribute is current.

6.6.2.1 Year
The screen is empty except for the year of the actual year.

6.6.2.2 Annual Calendar
The screen displays the overview calendar of the current year. With the following commands, the system remains on the annual calendar:

- ▼: The actual day is increased by one day, in combination with ⇑ or Ctrl by 1 week and with Alt by 1 year.
- Page↓: The actual day is increased by 1 month, in combination with ⇑, Ctrl or Alt by 1 year.
- ►: The actual day is increased by 1 week and in combination with ⇑ and with Alt by 1 month.
- ▲, ◄, Page↑: Analogous to ▼, Page↓ or ►, but in reverse time.
- Pos1: In combination with ⇑ or Ctrl, the actual day is the Monday of the current week, in combination with Alt it's the first of the actual month.
- End: Analogous to Pos1, with Sunday or end of the month.
- ⇆: The current day attribute cycles forward, in combination with ⇑ backwards.
- F2: continue with Search.

- F5: If a day attribute is actual, the actual day has this attribute.
- F6: If a day attribute is actual, the actual day has lost this attribute, if it applies to it. Afterwards, the actual day is increased by 1 day.
- Print: The annual calendar is printed.

In this case, the actual week, the actual month, or the actual year is always adjusted, possibly with a new issue. If an undefined date would arise (such as February 29 in a non-leap year or September 31), the last day of the actual month becomes the new actual day; if the range of defined calendar dates were left, nothing is changed.

6.6.2.3 Monthly and Weekly Calendars

The screen displays the overview calendar of the actual month or the actual week. The operation is completely analogous to the annual calendar.

6.6.2.4 Calendar Page

The screen displays the calendar page of the actual day.

If the calendar page contains no appointments, continue with Time, otherwise no cursor is visible, but the system is waiting for an input.

After entering ⏎ the cursor blinks in the field for the time of the first empty appointment on this day (if there is no more empty appointment, of the last appointment). The input of Esc leads to the screen changing to the weekly calendar of the current week; the cursor blinks in the field for the current day.

With ↑ or ↓ the current day is the next or previous one from the appointment calendar (i.e., on which the calendar page contains appointments). If the last r first day with appointments was already reached, the actual day remains unchanged.

With Print, the calendar page is printed.

6.6.2.5 Time, Appointment Attribute, Keyword, Text

The cursor blinks in the field of the time, the appointment attribute, the keyword, or text of the actual appointment; the corresponding object can be edited.

From the first of these components of an appointment to the third, you can get to the next one with and from the text to the time of the next appointment in the sequence (with ⇑⏎ you land there immediately).

With ⇑+▼ or ⇑+▲ the next or previous appointment becomes the actual appointment (if that was already the last or first one, nothing is changed); with ⇑Pos1 or ⇑End you land at the first or last appointment of the sequence.

The clipboard is used for moving and copying: With F7 the actual appointment is copied into it and removed from the appointment sequence, with F8 the same happens, but without deletion of the appointment, and with F9, the appointments from the clipboard are copied behind the current appointment, if there is still enough space on the calendar page.

The deletion of the current appointment is achieved with ⇧Del.

Esc leads back to the same day in the weekly calendar.

6.6.2.6 Search

The cursor blinks in the field of the search word, it can be edited. The completion of the input with ⏎ results in the calendar data of the days on which the entered search word appears in the keywords of any appointments being colour-marked in the calendar. The markings remain when the calendar is changed. An input completion with Esc cancels the search.

In both cases, you are then back at the place from which the search was called.

6.7 Construction

We only show the specifications of the packages.

6.7.1 Term Attributes

```
package attr
import . "µU/obj"

const Wd = 3
type Attr = uint

type Attribute interface {
  Object
  Editor
  Printer
  Stringer
  Marker
}
type
  AttrSet interface { // a set of appointment attributs

  Object

// a is inserted into the set.
  Ins (a Attribute)

// The appointment attributes from the set are displayed on the
// screen in short format in a string starting from position (1, c).
// If b == true, at position (1, c-1) a red space is output
// from position (1, c-1).
                  Write (1, c uint, b bool)
}
func NewSet() AttrSet { return newSet() }
```

6.7.2 Keywords

```
package word
import . "µU/obj"

const Wd = 12

type Word interface {

  Object
  Editor
  Stringer
  Printer

// Returns true, iff tha actual search term is contained in x.
  Ok() bool
}

func New() Word { return new_() }
```

6.7.3 Appointments

```
package appt
import (. "µU/obj"; "todo/attr")

const ( // format
  Long = Format(iota) // one complete screen line
  Short               // one line with 9 columns
)
type Appointment interface {
  Object
  Formatter
  Stringer
  Editor
  Printer

// Returns true, iff the actual search term
// is contained in the search word of x.
  HasWord() bool

// Returns the appointment attribute of x.
  Attrib() attr.Attribute
}
```

6.7.4 Appointment Sequences

```
package appts
import (. "µU/obj"; "µU/day")

type Appointments interface {
  Object
  Editor
  Printer

  SetFormat (p day.Period)

// Returns true, iff the actual search term is contained
// in one of the appointments in x.
  HasWord() bool
}
```

6.7 Construction

If the calendar page contains no appointments, continue with Time, otherwise no cursor is visible, but the system is waiting for an input.

After entering ⏎ the cursor blinks in the field for the time of the first empty appointment on this day (if there is no more empty appointment, of the last appointment). The input of Esc leads to the screen changing to the weekly calendar of the current week; the cursor blinks in the field for the current day.

With ▲ or ▼ the current day is the next or previous one from the appointment calendar (i.e., on which the calendar page contains appointments). If the last r first day with appointments was already reached, the actual day remains unchanged.

With Print, the calendar page is printed.

6.7.5 Persistent Sets of Calendar Data

```
package pdays
import (. "µU/obj"; "µU/day")

type PersistentDays interface { // persistent sets of calendardays

  Clearer
  Persistor

// Returns true, iff d is contained in x.
  Ex (d day.Calendarday) bool

// d is contained in x.
  Ins (d day.Calendarday)

// d is not contained in x.
  Del (d day.Calendarday)

// Returns the number of days in x.
  Num() uint
}
```

6.7.6 Day Attributes

```
package dayattr
import "µU/day"

// Manages a set of appointment attributes, one of which is actual,
// and for each of the attributes the persistent set of those days
// that have this attribute. In the beginning the first attribute
// in the file "dayattribute.cfg" is the actual attribute and
// for each attribute the persistent set of the days that have it
// diejenige, die es beim vorigen Aufruf des Programms war.

// The actual attribute is the first in the file "dayattribute.cfg".
func Normalize() { normalize() }

// For w == true the actual attribute is advanced by one in the cyclic
// sequence of the attributes and für w == false set back by one.
func Change (w bool) { change(w) }

// The actual attribute is written to the screen
// at position (line, column) = (l, c).
func WriteActual (l, c uint) { writeActual(l,c) }
```

```
// For b == true the actual attribute applies to d, not for
// b == false; the set of days that have it is corespondingly changed.
func Actualize (d day.Calendarday, b bool) { actualize (d,b) }

// The set of days that have the first attribute
// from the file "dayattribute.cfg" is empty.
func Clr() { clr() }

// Colour and font of d are set - depending on whether d is
// contained in the set of days that have the actual attribute
// or not and whether d is a holiday or not.
func Attrib (d day.Calendarday) { attrib(d) }

// For each attribute the set of days that have this one
// is secured in the file "dayattribute.cfg".
func Fin() { fin() }
```

6.8 Calendar Pages

```
package page
import (. "µU/obj"; "µU/day")

type Page interface {

  Object
  Editor
  Printer
  Indexer

  SetFormat (p day.Period)

// d is the date of x.
  Set (d day.Calendarday)

// Returns the date of x.
  Day() day.CalendardayG

// Returns true, iff the actual search term is contained
// in the keyword of an appointment of x.
  HasWord() bool

// s. dayattr.
  Fin()
}
```

6.8.1 Appointment Calendars

```
package cal
import "µU/day"

func SetFormat (p day.Period) { setFormat(p) }

// The actual calendar page is that of the day d.
func Seek (d day.Calendarday) { seek(d) }

// The appointments in the weekly calendar and all appointment
// attributes are written to the screen.
func WriteDay (l, c uint) { writeDay(l,c) }
```

6.8 Calendar Pages

```
// The sequence of calendar pages is changed by editing
// wobei bei d begonnen wird. d is danach
// the date of the last edited calendar page.
func Edit (d day.Calendarday, l, c uint) { edit (d,l,c) }

// The actual search term is the one that was edited at position (l, c).
func EditWord (l, c uint) { editWord(l,c) }

// The actual calendar page is printed,
// starting from (line, column) == (l, c).
func Print (l, c uint) { print(l,c) }

// The actual calendar page is persistently secured.
func Fin() { fin() }
```

Life

7

Life is life-threatening.

Ulrich Scholtze
Bournemouth, August 1967

Abstract

This project presents two *games*: the *Game of Life* by John Conway and a *predator-prey* system with foxes, hares and plants.

This teaching project was originally only designed for the simulation of a simple *predator-prey* system.

During the system analysis, something astonishing was discovered, which led to the project having "two faces".

7.1 System Analysis

The basic ideas of simulating a predator-prey system are highly related to those from the *Game of Life* by John Conway (in essence, both are a simple *cellular automaton*).

Consequently, the task consists of two parts:

- the *Game of Life* by John Conway and
- the simulation of a simple *predator-prey system*.

The consequence of this is the extensive bundling of the different manifestations in the representation of the data and the construction of the algorithms—not least with a view to easy expandability; furthermore, the possibility of choosing between the two simulations at the start of the program.

7.1.1 The Game of Life

In the *Game of Life*, there is only one type of creature: *cells*.
They survive according to the following rules:

- If they have more than three cells in the neighbourhood, they die of stress.
- If they do not have at least two cells in the neighbourhood, they die of loneliness.
- In a free space, where there are three cells in the neighbourhood, a new cell is created.

The up to eight adjacent spaces, not only horizontally and vertically, but also diagonally, are considered as *neighbourhood*.

There is a lot of literature on this topic (s. [1, 2, 4, 6]) and two websites that deal intensively with this topic and provide many references to literature (see [3, 5]).

7.1.2 The Ecosystem of Foxes, Rabbits, and Plants

foxes eat *hares*, *hares* eat *plants*. Both groups of creatures can only survive if their environment is not overpopulated with their own kind and they therefore can't find anything to eat.

The *ecosystem* is modelled as a rectangular world of checkerboard-arranged spaces. Each space is either occupied by a plant, a hare, or a fox. The *survival rules* are very simple:

- The space of a plant is taken over by a hare if there is a hare on one, two, or three neighbouring spaces ("hares eat plants").
- A hare loses its space to a plant if there are already four hares in the neighbourhood ("hare finds nothing to eat").
- The space of a hare is taken over by a fox if there is at least one fox on a neighbouring space ("foxes eat hares").
- A fox has to give up its space to a plant if there is no hare on any neighbouring space ("fox finds nothing to eat").

In the course of the simulation of the "generational" development of the ecosystem, the following is possible:

- A world initially only occupied with plants is "created" by the user (i.e., some of its spaces are occupied with hares and foxes).
- The development of a world according to the above rules is followed step by step (where the rules are applied once in each step).
- The simulation can be stopped at any time, the interim status archived, and can be restored and continued at any time.

Fig. 7.1 Component hierarchy of the Game of Life

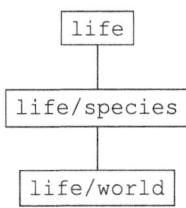

- Each world can be given a name.

The actual state of the world is clearly displayed on the screen.

7.1.3 The Objects of the System

The following objects can be derived from the system analysis:

- the different *life forms* in their places within the considered system and
- the *worlds*, in which they live.

Both are realized as abstract data types; the corresponding packages are life with the subpackages life/species and life/world.

7.1.4 Component Hierarchy

The dependencies of the packages are shown in Fig. 7.1, where the lower package is used (imported) by the one above it:

Furthermore, some project independent packages from the microuniverse are used, as, e.g., kbd, col, scr, and pseq.

7.2 User Manual

The life forms are represented pixel by pixel by small "icons" in the 16×16 grid.

For the screen size, PAL (768×578) is chosen; thus, the world is 48 spaces wide and 34 spaces high at an icon size of 16×16 pixels, with each space able to accommodate one creature of a species. The first line is reserved for the system's heading, and the last line is reserved for the system's operating instructions and error messages.

7.2.1 Program Operation

Since there are two different systems, it is initially determined which of the two should be called:

7.2.1.1 System Selection
After calling the program, a menu appears in which you can choose between

- the *Game of Life* or
- the *Ecosystem of Foxes, Rabbits, and Plants*.

When the input is completed with the enter key ↵, the corresponding system is selected, and the program is terminated with the escape key Esc.

7.2.1.2 World Definition
The cursor is in the field for the world's name; the field is empty, the name must be entered. If the name is empty or the input is completed with Esc, the program is terminated.

7.2.1.3 World Editor
If a world with the entered name already exists, it is loaded; a new world in the ecosystem is only full of plants, in the Game of Life it is empty.

The only keyboard inputs possible are ↵ and Esc; with ↵ a step of the simulation is performed according to the rules, and with Esc you return to the world definition.

The occupation of each space by a life form can be changed. In the *Game of Life*, a cell is inserted with a click of the left mouse button and removed with the right mouse button; in the *Ecosystem*, a hare is inserted with a click of the left mouse button and a fox in combination with the shift key, and a plant with the right mouse button.

Figure 7.2 shows the screen when the Game of Life was selected, with the "gun" — a figure that constantly "shoots" the same cell combinations. This world is named gun. Figure 7.2 shows an ecosystem.

Figure 7.3 shows an ecosystem.

7.3 Construction

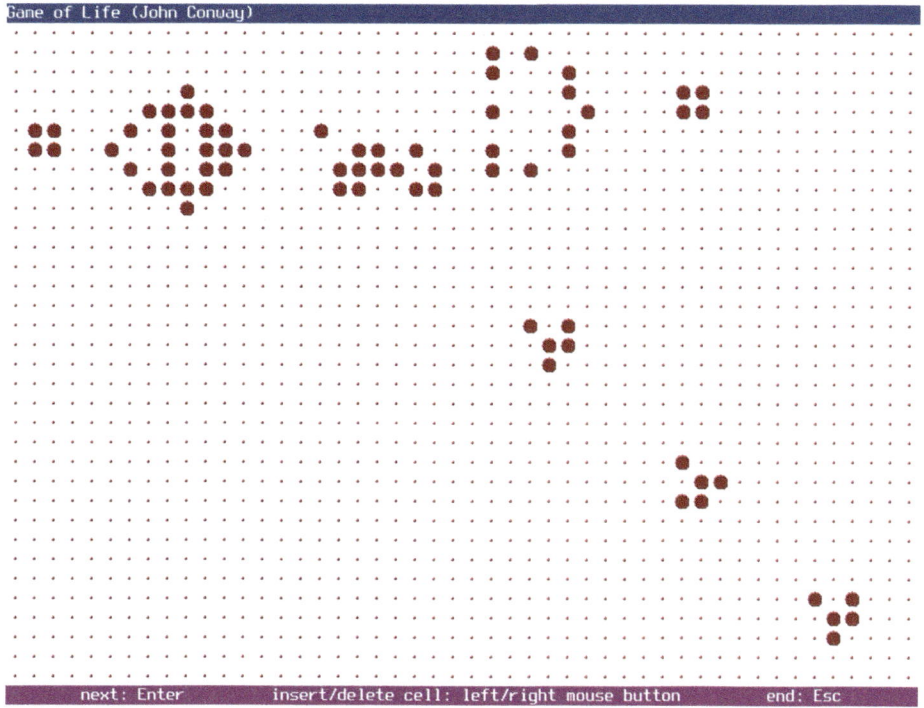

Fig. 7.2 The Game of Life: the gun

7.3 Construction

7.3.1 Specifications

Here is the specification of the lifestyle package:

```
package species
import . "µU/obj"

type System byte
const (
  Eco = System(iota) // Ecosystem with foxes, hares and plants
  Life               // Game of Life (John Conway)
)
var (
  Suffix string
  NNeighbours uint
)
type
  Species interface {

  Equaler
  Stringer

  Write (l, c uint)
```

Fig. 7.3 An ecosystem

```
// if k == 0 in Eco: x is a plant
//           in Life: x is nothing
// if k == 2 in Eco: x is a hare
//           in Life: x is a cell
// if k == 3 in Eco: x is a fox
//           in Life: x is a cell
  Set (k uint)

// The actual species has changed according to func.
  Modify (func (Species) uint)
}

// Returns a new species.
func New() Species { return new_() }

// The actual system is s.
func Sys (s System) { sys(s) }
```

and here is that of the world package:

```
package world
import (. "µU/obj"; "µU/mode"; "life/species")

const Len = 8 // maximal length of the name of the world
type World interface {
  Equaler
  Write()
```

7.3 Construction

```
  Edit()
  Stringer
  Persistor
}

// Returns a new empty world.
func New() World { return new_() }

// Returns the mode for life.
func Mode() mode.Mode { return m() }

// s is the actual system.
func Sys (s species.System) { sys(s) }
```

7.3.2 Implementations

Here are the representations of the abstract data types world and species

```
type species struct { byte }

type world struct {
  string "name of the world"
  spec, old []species.Species
  line, column uint16
}
```

We also show the main program:

```
package main
import ("µU/str"; "µU/col"; "µU/scr"; "µU/box"; "µU/errh";
        "µU/files"; . "µU/menue" "life/species"; "life/world")

var m = world.Mode()

func defined() (string, bool) {
  bx := box.New()
  w := scr.NColumns()
  bx.Wd (w)
  bx.Colours (col.Black(), col.FlashWhite())
  bx.Write ("world:" + str.New(w - 6), scr.NLines() - 1, 0)
  const n = world.Len
  bx.Wd (n)
  name := str.New (n)
  bx.Colours (col.FlashWhite(), col.Black())
  for {
    bx.Edit (&name, scr.NLines() - 1, 6)
    if str.Alphanumeric (name) {
      break
    } else {
      errh.Error0 ("Only letters and digits may appear in names")
    }
  }
  str.OffSpc (&name)
  errh.DelHint()
  return name, ! str.Empty (name)
}

func sim() {
  w := world.New()
  for {
    if name, ok := defined(); ok {
      w.Name (name)
```

```
       w.Write()
       w.Edit()
    } else {
       break
    }
  }
}

func main() {
  scr.New (0, 0, m)
  scr.ScrColourB (col.FlashWhite())
  scr.Cls()
  files.Cds()
  var x Menue
  x = New ("Game of Life")
  game := New ("Game of Life (John Conway)")
  game.Leaf (func() { world.Sys (species.Life); sim() }, true)
  x.Ins (game)
  var ecosys Menue
  ecosys = New ("Ecosystem of foxes, hares and plants")
  ecosys.Leaf (func() { world.Sys (species.Eco); sim() }, true)
  x.Ins (ecosys)
  x.Exec()
  scr.Fin()
}
```

References

1. Bell, D. I.: Spaceships in Conway's Life. Series of articles posted on comp.theory.cell-automata, Aug–Oct 1992. http://members.tip.net.au/~dbell
2. Buckingham, D. J., Callahan, P. B.: Tight bounds on periodic cell configurations in life. Exp. Mat. **7**:3, 221–241 (1998). https://www.emis.de/journals/EM/expmath/volumes/7/7.html
3. https://conway.life.com
4. Gosper, R. Wm.: Exploiting regularities in large cellular spaces. Physica D 75–80 (1984). https://doi.org/10.1016/0167-2789(84)90251-3
5. https://playgameoflife.com
6. Poundstone, W.: The Recursive Universe. MacGraw-Hill Contemporary (1985). ISBN 978-0809252022

The Go Register Machine

Make things as simple as possible,
but not simpler.

Albert Einstein

Abstract

The register machine is a machine model that is suitable for introducing the concept of computability. Its concept is equivalent to the *Turing Machine*.

At a teacher training conference at the Institute for Computer Science of the Free University of Berlin many years ago, Prof. Dr. S. Koppelberg gave a lecture on the question *What can algorithms do?* (see [2]). The central message was the mathematical equivalence of the concepts of the *Turing machine* and the *register machine* with respect to the computability of functions, which exactly characterizes the *recursive* functions. In this context, the article [1] is also pointed out.

The particular appeal of the topic lies in the contrast between the simplicity of the "programming language" of the register machine and the intellectual challenge of many tasks that can be solved with it—arbitrarily demanding tasks. Practical work with register machines is very simple using the Go programming language. Everything worth knowing about this is shown in this chapter.

The Go register machine is not a project from teacher training, but a program I constructed for use in teacher training. Therefore, this chapter does not include references to the phases of the program life cycle; nevertheless, its inclusion in this book is justified due to the importance of the concept for the fundamentals of computer science.

8.1 System Analysis

With register machine programs, basic concepts of both *imperative* (especially *machine-oriented*) and *functional* programming can be developed in a very natural way, because they model the structure of von Neumann computers quite well and also allow a "programming style" according to the functional paradigm.

The importance of this concept lies

- *Theoretically* in its equivalence to the concept of computability (by recursive functions or Turing machines): *Everything that can be programmed at all can in principle be done with register machines.*
- *Practically* in its clarity and comprehensibility: The necessary syntactic effort to construct RM programs is very low, therefore they are easier to handle than Turing machine programs that achieve the same.

Register machines thus represent an advantageous alternative to the introduction of Turing machines—especially in introductory considerations of computer science, because their treatment is possible without prior knowledge.

8.1.1 Components of a Register Machine

A register machine conceptually has

- a *data storage* in the form of a countable set of *registers* and
- a *program storage* for storing a program in the form of individual *program steps*.

The data storage consists of *registers*. They are memory cells that can be accessed directly. No distinction is made between the memory cells in the working memory and those in the processor.

Registers can hold a natural number as content. The content of a register is also referred to as its *value*. Since these values change during the execution of a program (which is precisely the purpose of RM programs), the registers can also be considered as variables.

Direct access to the registers is made possible by giving them names by which they can be addressed to access their values, e.g., $a, b, c, \ldots, x1, \ldots$.

Initially, all registers have the value 0.

Of course, the concept is an idealization insofar as on a real computer, due to the *finiteness* of its memory, neither can there be infinitely many registers nor can their values become arbitrarily large.

8.1.2 Basics of the Register Machine Programming Language

Register machines only have a very limited programming language with only five types of *instructions*:

- for the *assignment* of a value to a register, especially for creating a new register with the content 0;
- for the *modification* of a register content by +1 or −1;
- for the *jump* to another location in an RM program, also under the condition, that the value of a register is greater than 0;
- for the *return of a value* of a register or a function; and
- for the *output* of the value of a register or a function.

A register machine offers the possibilities

- for inputting a program, i.e., for populating the program memory with program steps and
- for outputting the intermediate and final results of calculations.

An RM program is executed by a register machine in such a way that its program lines are processed *sequentially* (line by line), starting with the first line and then either moving to the next program line or jumping to a line other than the next one.

Looking at it a bit more closely, this is realized as follows:

The next program line to be executed is always *that* one whose number is in a special register, the *program counter*. (However, the above-mentioned instructions cannot be applied to this register.)

Initially, the program counter contains a 0; so it starts with the first program line. Which number *after* the execution of a program line is in the program counter, i.e., which line is executed next, depends on the content of this program line.

If the value of the program counter is greater than or equal to the number of program lines, the execution of the program is finished.

Empty program lines only have the effect that the program counter is increased by 1.

8.1.3 System Architecture

We present here the implementation of the concept "register machine" as it is possible using the programming language Go.

8.1.4 Registers

The Go register machine uses the data type `Register`, which—along with the accesses to it—is encapsulated in the Go package μU from the microuniverse, which provides these things via an *interface*. This prevents access to the Go register machine that is *"unauthorized"* in the sense that it bypasses the syntax of the Go version of the RM language presented here.

Before the first use of a register, the Go register machine must be informed about its use, which in Go is done by creating an "object" of type `Register` with the value 0.

The size of the numbers that a register can contain as a value is limited by the range of the data type `uint` in Go (on a 64-bit computer up to $2^{64} - 1$).

8.1.5 Register Machine Programs

A Go register machine program—hereinafter briefly referred to as RM program—is implemented in Go as a *program package*. Its package header is *Go-specific*:

```
package main
import . reg
```

In the second line, the RM program is made aware of the type `Register` and its methods from the Go package `reg` for use.

We first consider the main function of a package body. It is enclosed in its signature `func main() {` and a closing curly bracket `}`.

The individual steps of the main function are placed *one after the other* in the program memory after their translation "line by line" and can therefore also be referred to as *program lines*. Each program line contains exactly one *instruction*; the main function is therefore a *sequence of instructions*.

The structure of an RM program in EBNF notation thus looks like this:

RMProgram = PackHead ";" PackBody.
PackHead = "`package main`" ";" "`import .reg`".
PackBody = [Funcs] FuncMain.
Funcs = Func ";" { Func ";" }.
FuncMain = "`func main`" Stmts "`}`".
Stmts = Stmt ";" { Stmt ";" }.

The literal ";" should be replaced by a line feed for better clarity of a program text. Therefore, it is also legitimate to refer to the individual program steps as "program lines".

The additional possibilities for constructing and using functions with "Func" are explained in the section on *functions*, those with "Stmt" in the following section.

8.1.6 Instructions

An instruction is

- a *value assignment*,
- a *change instruction*,
- a *jump instruction*,
- a *return instruction*, or
- an *output instruction*.

At the beginning of an instruction, a *mark* may additionally stand as a "jump target". There may also be "empty" instructions that consist *only* of a mark, or simply—for the sake of clarity—blank lines.

Here is the formal definition of an instruction in EBNF:

Stmt = [Label ":"] [Assign | IncOrDec | Jump | Return | Write] .
Label = CapLetter .
CapLetter = "A" | "B" | ...| "Z" | "_" | "Ä" | "Ö" | "Ü" .

Each of the individual instructions is dedicated to its own section below.

8.1.6.1 Value Assignment

The identifier for an *atomic register*—always a in the following descriptions—can be replaced by any string of letters and numbers that starts with a lowercase letter (in this sense, the a serves as a kind of "template"). The formal definition is given below in EBNF notation.

```
// a is a new register with the value 0.
  a := Null()
```

The second form of value assignment

```
// Pre: f is a registervalued function.
// a is a new register with this value.
  a := f()
```

for an *arbitrary* register-valued function f, we will go into more detail in the section on functions, because "RegValue" and "Register" in the following EBNF definition of value assignment also refer to the result values of functions.

Assign =   AtomicRegister ":=" RegValue .
AtomicRegister = SmallIdentifier .
RegValue =   "Null()" | FuncName [*Registers*] .
Register =   AtomicRegister | RegValue .
Registers =   Register { "," Register } .
FuncName =   Identifier .
Identifier =   Letter { Letter | Digit | "_" } .
SmallIdentifier = SmallLetter { Letter | Digit | "_" } .

Letter =	CapLetter | SmallLetter .
Digit =	"0" | "1" | ...| "9" .
CapLetter =	"A" | "B" | ...| "Z" | "Ä" | "Ö" | "Ü" .
SmallLetter =	"a" | "b" | ...| "z" | "ä" | "ö" | "ü" | "ß" .

Because we will also allow functions with results of type `bool`, the mentioned context-dependent condition that the function with the name "FuncName" delivers a value of type Register is of course essential.

There is a good reason why in the Go-RM language on the right side of a value assignment *no atomic register*, but only a function value may stand (`Null()` is one!).

Let's assume that the value assignment `a := b` would be allowed for an atomic register b (unfortunately, the Go translator would accept such a statement). Then the program

```
func main() {
  b := Null()
  b.Inc()
  a := b
  b.Dec()
  a.Write()
}
```

does not output 1, but 0! This is because the instruction `b.Dec()` also changes the value of the register a, i.e., with `a := b` not what was "intended" has been achieved, namely, to create a new register with the value of b, because its value is also changed by modification instructions to the register b, thus not representing its own register.

The "intention" is achieved with the instruction `a := Copy(b)`.

The deeper reason is that in the implementation of the Go register machine, the variables of type `Register` are only "references" to objects "behind the scenes" (i.e., pointers to addresses in memory where the objects are stored), not the objects "themselves". Consequently, with the instruction `a := b` only the start address of the register b would be copied into the start address of the register a, which means that the reference a points to the same object in memory as the reference b. This explains the "penetration" of the change from b to a.

8.1.6.2 Change Instruction

There are two instructions that can increase or decrease the value of a register; in EBNF notation: IncOrDec = AtomicRegister [".Inc()" \|".Dec()"].

Here are the details from the specification of the package reg:

```
// Pre: a was generated by a value assignment
// The value of a is incremented by 1.
  a.Inc()

// Pre: a was generated by a value assignment
//      and the value of a is bigger than 0.
// The value of a is decremented by 1.
  a.Dec()
```

8.1 System Analysis

In an RM program, it is therefore necessary to ensure the precondition, i.e., to ensure that the value of a is greater than 0 when the instruction `a.Dec()` is called. If this condition is not met, a program run will be aborted with a corresponding error message!

The test for this is provided by the method `Gt0` ("greater than 0") from the package `reg`, which returns a value of type *bool* (i.e., *true* or *false*):

```
// Returns true, iff the value of a is bigger than 0.
   a.Gt0()
```

The details on jump instructions follow in the next section.

The relationship between *precondition* and *effect* represents a kind of "contract" about the mutual *rights* and *obligations* of the programming person and the register machine:

The register machine has the *right*, to rely on the *precondition* that the value of x is greater than 0 when it is to execute the instruction `x.Dec()`, and for this the *obligation*, to effect the *effect*, i.e., to decrease the value of x by 1; the programming person has the *obligation*, to ensure that the *precondition* is met when the instruction is called, and thus the *right*, that the register machine effects the *effect*.

Why the change instructions are not allowed for registers in general, i.e., also for register function values, but only for atomic registers, becomes apparent from the following "program", which is (unfortunately) accepted (i.e., *translated*) by the Go translator, but does *not* deliver the "expected" result:

```
Two().Inc()
Two().Write()  // Oops ...
```

With the use of `Plus1()`, something adequate can be achieved:

```
Plus1r(Two()).Write()
```

8.1.6.3 Jump Instruction

Here we first only deal with the first alternative of the condition `Condition`; we will return to the second in the section on functions.

```
// If the value of a is bigger than 0, the program jumps to the program line
// with the mark A. Otherwise, the program is continued with the following,
// if there is one, otherwise, the program is terminated.
  if a.Gt0() { goto A }
```

If there is no program line with the label A, or if this label appears more than once, the program cannot be translated and the translation is aborted with a corresponding error message.

The same applies to the name of the label A as to that of a register.

Since the Boolean constant `true` can also be used for the condition, we also have the *unconditional* jump

```
  if true { goto A }
```

or just briefly

```
  goto A
```

The condition is used in the syntax for the *conditional jump*:

Jump =   "if" Condition "{" "goto" Label "}" .
Condition =   Register "." "Gt0()" | BoolExpression .
BoolConst =   true | false .
BoolExpression =   BollConst | BoolValue .
BoolValue =   FuncName [*Registers*] .

Here too, a context-dependent condition must be met: The function named "FuncName" returns a value of type bool.

8.1.6.4 Return Instruction

This instruction consists of the keyword return and a register or a Boolean value ...:

```
return ...
```

More details will follow in the next section.

8.1.6.5 Output Instruction

In addition to the instructions presented so far, our Go register machine contains the method Write from the package reg

- for the *output* of the value of a register or a function:

```
// The value of a is written to the screen (in a new line).
  a.Write()
```

8.1.7 Test Programs

a is either a register or the value of a register-valued function. Formally in EBNF: Write = Register."Write".

The results are checked in a Go program. The name of the file containing its source code must have the suffix .go; it is recommended to use main.go. The program is then translated with the command

```
go run main.go
```

bound to an executable program, and then called.

Of course, it cannot be ruled out that errors may occur during the translation. Possible causes include, for example, typing errors or non-compliance with the syntax described here.

If something like this happens, carefully check your source code—the error messages from the Go translator will certainly give you important clues as to what could be wrong.

Here is a concrete example:

8.1 System Analysis

```
package main
import . "µU/reg"
... (source texts of the functions Two and Four)
  func main() {
    x := Two()
    x.Print()
    Four().Print()
  }
```

It is very useful—if not necessary—to insert additional *comments* in the program text, which do not belong to the actual program text, but only serve to explain the program construction.

If this is done consistently in such a way that properties of the register values are described as a comment after each instruction, the *correctness* of an RM program can be proven. A "short form" is acceptable in which, for example, "the register a has the value 1" is abbreviated to "a = 1". We demonstrate this in the examples.

From a "software technical" point of view, it is also essential to provide a *specification* for each function in the form of a comment before its signature line, i.e., if its call depends on prerequisites, which ones, and what the function returns!

In Go—and thus in the Go-RM language—it is agreed:

All texts in program lines that follow two consecutive slashes // are not considered program text, but are ignored during translation.

8.1.8 Functions

Functions represent, in a way, RM ("sub-") programs that open up the possibility to significantly expand the scope of the RM ("programming") language. They return a register (and thus its value) or a Boolean truth value as a result.

Values of functions—such as the "number" Two()—can be considered as new "objects" of the RM language, which may be treated like a register in an RM program. In particular, they can also be assigned to an atomic register—such as Null() as a new register with the value 0—(which shows that Null() is basically also such a function).

8.1.8.1 Specification of Functions

Functions represent, in a way, RM ("sub-") programs that open up the possibility to significantly expand the scope of the RM ("programming") language. They deliver as a result either a register (and thus its value) or a Boolean truth value.

A *function* consists of the specification of its *signature*, its body—a sequence of program lines—and the *closing line*.

The first line of the program text of a function consists of its *signature*. It is introduced by the keyword func, followed by the name of the function (a string of letters and numbers that begins with a capital letter). This is followed by a pair of round brackets () and the keyword Register or bool. Within the brackets, one or—separated by commas—several register names may stand as *parameters*, followed by the keyword `Register`. An opening curly bracket { completes the signature.

The *body* of a function consists of a sequence of program lines, with which the value of the function is calculated, each line containing exactly one instruction.

Each "auxiliary register" a, i.e., one that does not appear as a parameter in the signature, must of course be created with a value assignment `a := ...` before its use (preferably at the beginning of the body).

The last instruction must be a *return statement* of the form `return y`, where `y` is the register or the Boolean expression whose value the function should deliver as a result.

Return statements may also be used in the middle of the function body, when the register or the Boolean expression is calculated; then the calculation is aborted and its value is delivered as a result.

The *closing line* consists only of a closing curly bracket {.

The whole thing a bit shorter in EBN notation:

Func =	RegFunc | BoolFunc.
RegFunc =	RegFuncSig RegBody RegReturn.
BoolFunc =	BoolFuncSig BoolBody BoolReturn .
RegFuncSig =	"func" FuncName "(" [Param] ")" "Register" "{".
BoolFuncSig =	"func" FuncName [*Param*] "bool" "{".
Param =	Identifier { "," Identifier } "Register".
RegBody =	[Stmts] RegReturn.
BoolBody =	[Stmts] BoolReturn.
End =	"}".

8.1.8.2 Are Function Values "Registers"?

Values of functions can be considered as new "objects" of the RM language, which can be treated *almost* like registers in an RM program; they are also stored in registers.

If `y` is such a function value, then

- *Assignments to a register* (`a := y`),
- *Jump instructions* (`if y.Gt0() { goto A }`),
- *Return instructions* (`return y`), and
- *Output instructions* (`y.Write()`).

Patterns for permissible instructions. The creation of a new register with the value 0 is a special case of value assignment: the function `Null` from the package `reg` is simply a (internally defined) function that delivers a register with the value 0.

The *modification instructions*, however, form an *exception*: program lines of the form

- `y.Inc()` or `y.Dec()`

are meaningless if `y` is a function value; the Go translator would, for example, respond to the program line

```
Null().Inc().Write()
```

with an error message.

Here is a minimal example in which everything mentioned occurs:

```
func null() Register {
  return Null() // return value
}
func main() {
  null.Write() // output
  if null().Gt0() { goto A } // jump
  n := null() // value assignment
  n.Write()
A:
  null.Write() // does not happen
}
```

8.2 User Manual

Go-RM programs are ultimately Go programs and thus subject to their syntactic requirements.

A program named `main.go` is then translated with the command "`go run main.go`", bound to an executable program and called. If errors appear during the translation, carefully check your source code—the error messages from the Go translator give you hints on what could be wrong.

It is very useful—if not necessary—to insert additional *comments* in the program text, which do not belong to the actual program text, but only have the meaning of explaining the program construction.

If this is done consistently in such a way that properties of the register values are described as comments after each instruction, the *correctness* of an RM program can be proven. A "short form" is acceptable in which, for example, "the register tt a has the value 1" can/should be abbreviated with "a == 1". We demonstrate this in the examples.

From a software technical point of view, it is also indispensable to provide the *specification* of each function in the form of a comment before its signature, i.e., the prerequisites for its use, if there are any, as well as the indication of the result value it delivers. In Go—and thus in the Go-RM language—it is agreed that all texts in program lines that follow two consecutive slashes // are not considered as program text, but are ignored during translation.

8.2.1 Examples

Here is a simple example of a function that returns a register with the value 2:

```
func Twp() Register {
z := Null()
z.Inc()
z.Inc()
return z
```

We now present a more challenging example, the calculation of the sum of two registers, with which we introduce typical patterns in the construction of RM programs and point out a possible "standard" error.

The following attempt is a naive approach:

```
func Sum (a, b Register) Register {
  if b.Gt0() { goto A }
  return a
A:
a.Inc()
b.Dec()
if b.Gt0() { goto A }
return a
}
```

However, this approach is *not a solution* to the problem!

The function does return the correct result, which can be immediately confirmed by thinking about its source code and, for example, by calling

```
Sum (Two(), Four()).Write()
```

in a short RM program. However, as a side effect, it has set the value of the register passed as the first parameter to the sum of the original values of both registers and counted down the value of the register passed as the second parameter to 0. This fact is also immediately apparent and can be demonstrated with the lines

```
a := Two()
b := Four()
Sum (a, b).Write()
a.Write()
b.Write()
```

in an RM program: This gives the value 6 for a and the value 0 for b, not the values 2 and 4. This (completely unacceptable!) phenomenon occurs in all such constructions.

To prevent the values of the registers passed as parameters from being changed, it must be ensured that their values match the original values at the end of the function call. This can be most easily achieved as follows: At the beginning of a function body, "helper registers" are created as copies of the passed registers, which are then used for the calculations instead of the passed registers.

To this end, we construct a function that returns a copy of a register. In it, the decrementing of the value of the register a is "logged" by "counting along", which is used after the calculation of intermediate results to restore the original value by corresponding "counting up".

8.2 User Manual

```
func Copy (a Register) Register {
  b := Null() // register to record the copy
  if a.Gt0() { goto A }
  return b // a == b == 0
A: // let x be the value of a when called
  h := Null() // helpregister to record the number of a.Dec()-statements
// h == b == 0, a + h == a == x
B:
  a.Dec()
  b.Inc()
  h.Inc() // a + h == x, h == b > 0
  if a.Gt0() { goto B }
// a == 0, consequently a + h == x == h, deshalb b == x,
// but because x == h > 0 we have a < a + h == x
// however, a == x must apply to the return of b,
// therefore, increase a as often as is recorded in h:
C:
  a.Inc()
  h.Dec() // a + h == x
  if h.Gt0() { goto C }
// h == 0, consequently a == x
  return b // b == a
}
```

With the use of this function, we obtain a *correct* solution for calculating the sum of two registers:

```
func Sum (a, b Register) Register {
  a1 := Kopie (a)
  b1 := Kopie (b) // a1 + b1 == a + b
  if b1.Gt0() { goto A }
  return a1 // a1 == a and still b1 == 0
A:
  a1.Inc()
  b1.Dec() // a1 + 1 + b1 - 1 == a1 + b1 == a + b
  if b1.Gt0() { goto A }
  return a1 // b1 == 0, therefore a1 + b1 == a1 == a + b
}
```

Here is another example of a Boolean function, the examination of whether two registers have the same value. Initially, the values of the passed registers are copied into helper registers for the reason mentioned above.

```
func Equal (a, b Register) bool {
  a1 := Copy (a) // a1 == a
  b1 := Copy (b) // b1 == b, therefore, a + b1 == b + a1
A:
  if a1.Gt0() { goto B }
// a1 == 0
  if b1.Gt0() { goto F } // a1 = 0 < b1
  return true // b1 == 0 == a1, therefore, a == a + b1 == b + a1 == b
B: // a1 > 0
  if b1.Gt0() { goto C } // a1 > 0, b1 > 0
  return false // a1 > 0 == b1, therefore, a + b1 == a == b + a1 > b
C: // a1 > 0, b1 > 0
  a1.Dec()
  b1.Dec() // a + b1 == b + a1
  goto A
F:
  return false // a1 == 0 < b1, therefore a < a + b1 == b + a1 == b
}
```

A somewhat more complicated example with nested loops is the calculation of the sum of the first *n* natural numbers:

```
// Returns 0, if a has the value 0, otherwise the sum of the first n natural numbers,
// where n is the value of a.
func Gauß, (a Register) Register {
  g := Null()
  if a.Gt0() { goto A }
  return g // g == a == 0
A:
  a1 := Kopie(a)
  b := Null()
  c := Null()
B:
  a1.Dec()
  b.Inc()
C:
  g.Inc()
  c.Inc()
  b.Dec()
  if b.Gt0() { goto C }
D:
  c.Dec()
  b.Inc()
  if c.Gt0() { goto D }
  if a1.Gt0() { goto B }
  return g
}
```

The source code of this—albeit correct—example is definitely *bad*: It contains no comments on the values of the respective registers and no information on loop invariants; therefore, it is difficult to understand the algorithm, and the proof of its correctness is missing.

8.2.2 Recursion

However, the following fact is essential to the concept of "nesting":

In the instructions of an RM program—including in the body of functions—already *existing* functions can be used. In particular, it is possible to "nest" function calls (even multiple times).

A simple example of this is the following:

An example of possible multiple nesting is

```
func Six() Register {
return Times3(Two())
}
```

An example for the possible multiple nesting is

```
func Hundredninetytwo() Register {
  return Times3(Sum(Four(), Sum(Six(), Times3(Times3(Six())))))
}
```

with

```
func Times3 (a Register) Register {
return Sum (a, Sum (a, a)
}
```

This leads us to the idea of formulating algorithms recursively.

8.2 User Manual

The elegance of this approach can be easily demonstrated. We show a *significantly* simpler solution to the Gaussian problem, the correctness of which is immediately clear because the algorithm is precisely the definition of the result:

```
func Gauß (a Register) Register {
  if a.Gt0() { goto A }
  return Null()
A:
  b := Kopie (a)
  b.Dec()
  return Sum (a, Gauß (b))
}
```

Thus, algorithms in the Go-RM language can be formulated as elegantly as in *functional programming languages*. For example, all operations of arithmetic can be developed by recursion, as is common in mathematics.

These recursive versions are significantly easier to understand than those in section subsec:gormexamples.

Using the function

```
// Returns a register, whose value is 1 greater than the value of a.
func Succ (a Register) Register {
  b := Copy (a)
  b.Inc()
  return b
}
```

we show this using the example of the sum:

func Summe (a, b Register) Register if a.Gt0() goto A return Kopie (b) // a == 0, folglich a + b == b A: c := Kopie (a) c.Dec() return Succ (Summe(c, b))

This algorithm is correct because it precisely represents the definition of the sum from the theory of natural numbers. Also, the creation of a copy of a register and the checking for matching two register values can be implemented recursively: func Kopie (a Register) Register if a.Gt0() goto A return Null() A: a.Dec() b := Kopie (a) a.Inc() b.Inc() return b

func Equal (a, b Register) bool if a.Gt0() goto A if b.Gt0() goto F // a == 0 < b return true // a == b == 0 A: // a > 0 if b.Gt0() goto B return false // b == 0 B: a1 := Copy (a) a1.Dec() b1 := Copy (b) b1.Dec() return Equal (a1, b1) F: return false

8.2.2.1 Primitive Recursion

Basically, these examples are patterns for primitive recursion, which is very simple with the Go register machine:

For functions

```
func g (a Register) Register
func h (a, b, c Register) Register
```

we immediately get

```
func f (a, b Register) Register {
  if a.Gt0() { goto A }
  return g (b) // a == 0
A:
```

```
c := Copy(a)
c.Dec()  // c == a - 1
return h (c, b, f (c, b))
}
```

As a simple example, based on the sum function—itself an example of primitive recursion—here is the product formation:

If we substitute for tt g and tt h the functions

```
func g (a Register) Register { return Null() }
```

and

```
func h (a, b, c Register { return Summe (a, c) }
```

f(a, b) have the value of the product of the values of a and b.

8.2.2.2 μ-Recursion

For $f: \mathbb{N}^{k+1} \to \mathbb{N}$, the partial function $\mu f: \mathbb{N}^k \to \mathbb{N}$ is defined as follows:

$$\mu f(a_1, a_2, \ldots, a_k) = n \Leftrightarrow f(a, a_1, a_2, \ldots, a_k) = 0 \text{ und}$$

$$\text{for all } b < a \text{ gilt } f(b, a_1, a_2, \ldots, a_k) > 0$$

$$\mu f(a_1, a_2, \ldots, a_k) \text{ undefiniert} \Leftrightarrow \text{for all } b \in \mathbb{N} \text{ gilt } f(a, a_1, a_2, \ldots, a_k) > 0$$

This can be replicated—with the function types from the package μU using the register sequences:

```
func μ(f RegFunc1) RegFunc {
  return func (as Registers) Register {
           a := Null()
           goto B
        A:
           a.Inc()
        B:
           if f(a, as).Gt0() { goto A }
           return a
        }
}
```

The execution of this function does not terminate precisely when f(a, as).Gt0() for all registers Null(), One(), Two(), Three(), ... applies.

This makes it clear that the class of functions that can be calculated with Go-RM programs includes the *recursive* functions.

That the "initial functions" of the class of recursive functions—the constant functions, (especially Null), the successor function plus1, and the projections—can also be expressed by Go-RM functions, is sufficiently proven by the examples in the previous section, and the *substitution* of functions is a syntactic part of the Go-RM language—so everything is said about that.

Of course, the proof that *every* Go-RM function is *recursive* could also be led, closely following [2], Sects. 4. and 5, or [3], Sect. 2.4, by means of *Gödelization of the Go-RM functions*. However, a modification of the technical details of these proofs to the Go-RM

8.2 User Manual

functions can be omitted here with good conscience, as these are not new findings and it should be intuitively clear that the Go-RM functions are *by no means* more powerful than the *recursive* functions.

But it's simpler:

The reversal is also—referring to the equivalence of the classes of WHILE—computable programs and the μ-recursive functions—easy to show, because every Go-RM function can be easily transformed into the form of a WHILE program according to the procedure in Sect. 2.3 of [3].

We assume that the program contains neither lines that consist only of a mark (if necessary, such lines are combined with the one immediately following) nor those that contain several instructions separated by semicolons (if necessary, each semicolon is replaced by a line feed). The transformation is made as follows:

The program lines are numbered (starting with 0 for the function signature). From lines with a mark at the beginning, the mark and the colon following it are removed.

Jump instructions of the form

```
if a.Gt0() { goto A }
for n == z { n = next(a, z, n) }
```

are transformed into the form

```
for n == z { s; n++ }
```

where z is the number of the line that began with the mark A; those of the form goto A into the same form with a = One().

Other instructions s are converted to

```
for n == z { s; n++ }
```

transformed (z as above the line number). The following auxiliary function is used:

```
func next(a Register, z, n uint) uint {
  if a.Gt0() {
    return z
  }
  return n + 1
}
```

The sequence of lines resulting in the body of the function is enclosed with

```
n := 1
for n > 0 {
```

and

```
}
```

enclosed.

In addition to the Go-RM operations that make up the instructions, this method only uses syntactic components of Go that are allowed in WHILE programs: elementary handling of natural numbers—here of type int—and for loops (with the semantics of WHILE loops).

We demonstrate this with a simple example—the "translation" of the function

```
func Copy(a Register) Register
```

from Sect. 8.2.1 on program examples:

```
func Copy (a Register) Register {
  n := 1
  for n > 0 {
    for n == 1  { b := Null(); n++ }
    for n == 2  { n = next (a, 4, n) }
    for n == 3  { return b }
    for n == 4  { h := Null(); n++ }
    for n == 5  { a.Dec(); n++ }
    for n == 6  { b.Inc(); n++ }
    for n == 7  { h.Inc(); n++ }
    for n == 8  { n = next (a, 5, n) }
    for n == 9  { a.Inc(); n++ }
    for n == 10 { h.Dec(); n++ }
    for n == 11 { n = next (h, 9, n) }
    for n == 12 { return b }
  }
  return b
}
```

If, as in this case, only return instructions with the same register b as the result value occur, these lines can also be replaced by n = 0 if return b is added as the last program line.

This makes it clear that *every* Go-RM function is *recursive*.

8.2.2.3 Encoding Functions

For the proof network of the equivalence between Turing, GOTO and WHILE computability, bijective functions $N \to \mathbb{N}^k$ play an important role. They can also be developed in the Go-RM language.

We first consider the encoding function $c0 :: \mathbb{N} \times \mathbb{N} \to \mathbb{N}$, defined by

$$c_0(n, m) = \binom{n+m+1}{2} + n = \frac{1}{2}(n+m+1)(n+m).$$

In Table 8.1, the first function values of this encoding function are shown.

Because of $c0(n, m) = c0(0, n + m) + n$ the inverse function $d0 \colon \mathbb{N} \to \mathbb{N}^2$ of $c0$ can be easily found for decoding: It is defined by $d0(n) = (d(n), e(n))$ for

$$e(n) = n - c_0(0, y_0) \quad \text{und}$$
$$f(n) = y0 - (n - c_0(0, y_0)) \quad \text{mit}$$
$$y_0 = \max\{y \in \mathbb{N} \mid y \leq n \wedge c_0(0, y) \leq n\}.$$

This "encoding/decoding" principle can be generalized to any $k \in \mathbb{N}$ ($k \geq 2$): The k-digit encoding function $c \colon \mathbb{N}^k \to \mathbb{N}$, given by

$$c(n_0, n_1, \ldots, n_k) = c_0(n_0, c(n_1, \ldots, c_0(n_k, 0))),$$

provides a bijection $\mathbb{N}^k \to \mathbb{N}$.

Its inverse function $d \colon \mathbb{N} \to \mathbb{N}^k$ for decoding looks as follows:

8.2 User Manual

Table 8.1 The first function values of c_0

m ↓	$n \rightarrow$	0	1	2	3	4	5	6	...
0		0	2	5	9	14	20	27	...
1		1	4	8	13	19	26	34	...
2		3	7	12	18	25	33	42	...
3		6	11	17	24	32	41	51	...
4		10	16	23	31	40	50	61	...
5		15	22	30	39	49	60	72	...
6		21	29	38	48	59	71	84	...
⋮		⋮	⋮	⋮	⋮	⋮	⋮	⋮	

$$d(n) = (d_0(n), d_1(n), \ldots, d_k(n)) \text{ für } n \mathbb{N}^k$$

with $d_0(n) = e(n)$ and $d_{i+1}(n) = d_i(f(n))$ for $0 \leq i < k$.

The proof that d is inverse to c follows immediately by induction from the recursive definitions, based on the fact that $d0$ and $c0$ are inverses of each other.

All these functions can be implemented in the GO-RM language:

```
func C0 (a, b Register) Register {
  if a.Gt0() { goto A }
  return Null()
A: // a > 0
  c := Copy(a)
  c.Dec() // c == a - 1
  return Sum (C0 (c, b), a)
}

func D0 (n Register) (Register, Register) {
  return E(n), F(n)
}

func max (a Register) Register {
  b := Copy (a)
A:
  if Leq (C0 (Null(), b), a) { goto B }
  b.Dec()
  goto A
B:
  return b
}

func E (n Register) Register {
  y0 := max(n)
  return Difference (n, C0 (Null(), y0))
}

func F (n Register) Register {
  y0 := max(n)
  x := Difference (n, C0 (Null(), y0))
  return Difference (y0, x)
}
```

}

The functions `Leq` and `Difference` are left as an exercise.

```
func C (n ...Register) Register {
  if Equal (Length(a), Two) { goto A }
  return C0(n[0], C(n[1:]...))
A:
  return C0(n[0], n\cite{ch8F})
}

func d (i, n Register) Register {
  i1 := Copy (i)
  if i1.Gt0() { goto A }
  return F (n)
A:
  i1.Dec()
  return d (i1, n)
}

func D (i, n Register) Register {
  i1 := Copy (i)
  if i1.Gt0() { goto A }
  return E (n)
A:
  i1.Dec()
  if i1.Gt0() { goto B }
  return E (F (n))
B:
  return d (i1, F (n))
}
```

For this reason, the results from Chap. 2 of [3] can be fully transferred.

8.3 Construction

The register package is included in the microuniverse. Here is its specification:

```
package reg

type Register interface { // Registers with integer values.
  // For all methods, the preposition is that the calling
  // register was generated by a value assignment of Null()
  // or the value of a function with a register as value.
  // The calling register is always denoted by "x".

// Pre: The value of x is incremented by 1.
  Inc()

// Pre: x has a value > 0.
// The value of x is decremented by 1.
  Dec()

// Returns true, iff x has a value > 0.
  Gt0() bool

// Returns a new register with the sum
// of the values of a and x as value.
  Add (a Register) Register

// Returns a new register with the product
// of the values of a and x as value.
  Mul (a Register) Register
```

8.3 Construction

```
// The value of x is written in a line to the screen.
  Write()
}

// Returns a new register with the value 0.
  func Null() Register { return null() }
```

and here, exceptionally—because of its brevity—its implementation: package reg

```
type register struct {
  uint "value of the register"
}

func null() Register {
  return new(register)
}

func fail (s string) {
  panic ("Pre of " + s + "() not met")
}

func (a *register) Inc() {
  a.uint++
}

func (a *register) Dec() {
  if a.uint <= 0 { fail("Dec") }
  a.uint--
}

func (a *register) Gt0() bool {
  return a.uint > 0
}

func (a *register) Add (b Register) Register {
  c := null().(*register)
  c.uint = a.uint + b.(*register).uint
  return c
}

func (a *register) Mul (b Register) Register {
  c := null().(*register)
  c.uint = a.uint * b.(*register).uint
  return c
}

func (a *register) Write() {
  z := a.uint
  if z == 0 {
    println ("0")
    return
  }
  s := ""
  if z < 0 {
    s = "-"
    z = -z
  }
  n := z
  var t string
  for t = ""; n > 0; n /= 10 {
    t = string(n \% 10 + '0') + t
  }
  println (s + t)
}
```

8.4 Exercises

For the specifications of the tasks, let's agree on a somewhat "sloppy" way of speaking: We use the terms "register" and "value of the register" synonymously.

Implement the following functions and test them:

```
// Returns a - b, if a > b, otherwise 0.
func Difference (a, b Register) Register

// Returns a * a.
func Square (a Register) Register

// Returns a * b.
func Product (a, b Register) Register

// Returns the greatest register b with b * b <= a.
func Root (a) Register

// Returns 2$^a$.
func Exp2 (a Register) Register

// Returns a$^b$.
func Exp (a, b Register) Register

// Returns a / 2.
func Div2 (a Register) Register

// Pre: b > 0.
// Returns a / b.
func Div (a, b Register) Register

// Returns a \% 2.
func Mod2 (a Register) Register

// Returns a \% b.
func Mod (a, b Register) Register

// Pre: a > 0, b > 0.
// Returns the greatest common divisor of a and b.
func Gcd (a, b Register) Register

// Pre: a > 0, b > 0.
// Returns the smallest common multiple of a and b.
func Scm (a, b Register) Register

// Returns a!.
func Fakulty (a Register) Register

// Returns the maximum of a and b.
func Max (a, b Register) Register

// Returns the minimum of a and b.
func Min (a, b Register) Register

// Returns log$_2$(a).
func Log2 (a Register) Register
```

```
// Returns log$_b$(a).
func Log (a, b Register) Register

// Returns the binomial coefficient $\binom a b$.
func Binom (a, b Register) Register

// Returns die a-th Fibonacci-number.
func Fibonacci (a Register) Register

// Returns true, iff a < b.
func Less (a, b Register) bool
```

The book on the Mathematical Aspects of Applied Computer Science can be found on the worldwide web in the Monographs and Lecture Notes of the European Mathematical Information Service (EMIS):

http://www.emis.de/monographs/schulz/algo.pdf.

References

1. Fehr, E.: Mathematische Aspekte der Programmiersprachen. In: Schulz, R.-H. (Hrsg.) Mathematische Aspekte der Angewandten Informatik, S. 147–164. BI-Wissenschaftsverlag (1994)
2. Koppelberg, S.: Was können Algorithmen? In: Schulz, R.-H. (Hrsg.) Mathematische Aspekte der Angewandten Informatik, S. 23–54. BI-Wissenschaftsverlag (1994)
3. Schöning, U.: Theoretische Informatik – kurzgefasst, 3. Aufl. Spektrum Akademischer Verlag (1997)

The Electronic Stylus

9

Who is on the table today?

Typical question from teachers at the beginning of a lesson

Abstract

The electronic stylus is not a "project". It was originally just a test program for *sequences of objects* from the microuniverse, where later for the objects—out of annoyance over a completely nonsensical example from a book on object-oriented programming with Java—an abstract data type "two-dimensional figures" was chosen. The system is suitable for supplementing blackboard writing with computer and projector use.

One advantage of this—compared to the powerful system E-Chalk by Prof. Dr. R. Rojas from the Free University of Berlin—*tiny* system is the drastic reduction of the concept due to its "slimness": Its object-based construction and the event control in the main program are easily manageable.

9.1 System Analysis

The use of an electronic system can—under certain conditions—supplement or even replace writing and drawing on a board. The system essentially does the same as the "classic" board writing, but much more: The simulated board writing

- *can be designed very cleanly*,
- is *reproducible*,

- *modifiable and expandable* as well as
- *printable*.

The advantages and disadvantages compared to writing and drawing on a board must, of course, be carefully weighed against each other in each individual case—depending on the purpose of use; any mixed forms are conceivable.

Disadvantages

- Everyone can handle *chalk*; the operation of a *program*, on the other hand, must be learned.
- The students also need to learn how to operate it if they want or need to use the program.
- Freehand drawings or texts usually suffer more from a "shaky hand" when created by a mouse on an "electronic board" than when written on a board.
- The use is *dependent* on *devices* (computer, projector) that must be set up before use.
- There may be additional *costs* for the equipment, or a suitably equipped teaching room must be sought.
- The *follow-up costs* are also not negligible (replacement lamps for projectors, for example, are considerably more expensive than—even coloured—chalk).
- The program requires a computer (or network access to a server) under *Linux* (whether this is *really* a disadvantage …).

Advantages

- The program is very *easy to use*. Its control is easy to change and its ergonomic weaknesses or errors are easily correctable.
- The entire source code is freely usable for teaching situations; this ensures adaptability to individual needs.
- With a few "mouse clicks" it is possible to
  - draw clean figures, e.g., triangles, quadrilaterals, rectangles, circles, ellipses, and elegant curves;
  - write text cleanly; and
  - integrate images.
- Board wiping is *also* (in doubt: "dirty") work.
- The additional cost can be cushioned if a generally available "*mobile station*" is used.
- The program also runs on other "window-oriented" operating systems if the local computer has network access to a server under Linux.
- The user is *facing the audience* when using it (and does not, for example, catch paper balls from behind).
- An electronic board—let's call it "eBoard" for short—can be optimally *prepared* and *post-processed* and "*styled*" at any computer in peace.

9.1 System Analysis

- The eBoards are *reusable at will* ("… this is what our board looked like the day before yesterday").
- The long-term *maintenance* (i.e., the modification based on experiences, the adaptation to other teaching situations, and the further development) of the eBoards is possible at any time.
- The eBoards can be *passed on as files, printed out* and thus *duplicated*.

9.1.1 The Figures of the Electronic Stylus

The electronic stylus must be able to manage the following types of figures:

- *Point sequences,*
- *Line segments,*
- *Polygons,*
- *Curves,*
- *Straight lines,*
- *Rectangles,*
- *Circles,*
- *Ellipses,*
- *Texts,* and
- *Images.*

The following will explain these figures individually. But first, a few words about the colour concept: All figures can be displayed in different colours.

The *background colour* of the screen is black by default, but can be switched to white "at the push of a button".

Point sequences
These are *sequences of individual points* ("pixels" on the screen), of which every two consecutive ones, if they are not adjacent as pixels, are connected by a line segment.

The points are generated independently (i.e., by the program during movement) by the movement of a pointing instrument (currently only the mouse; perhaps a graphics tablet in the future) until the user input is terminated (the order of the points is given by the temporal sequence during generation).

Point sequences thus realize "freehand drawings" (which can of course also represent texts).

Line segments

Line segments are sequences of lines, where the endpoint of each line coincides with the starting point of the next line, provided there is a following one (the order of the lines is given by the temporal sequence of setting their endpoints).

In principle, line segments are the same as point sequences; the difference is that the boundary points of the lines are set individually by the user instead of being generated by mouse movement.

Polygons

Polygons are "closed" line segments: The starting point of their first line coincides with the endpoint of their last one.

Thus, *triangles*, *quadrilaterals*, *pentagons*, etc. belong to the figures.

Curves

Curves are defined by Bezier polynomials, which are defined by—individually set by the user—point sequences, limited by a certain maximum number.

A curve has the first and last point of the sequence as boundary points; the points in between are the "support points" of the polynomial, whose degree n is 1 less than the number of points:

$$z(t) = \sum_{i=0}^{n} \binom{n}{i}(1-t)^{n-i}t^i z_i \text{ for } 0 \leq t \leq 1 \text{ und } z_0, z_1, \ldots, z_n \in \mathbb{C}$$

Straight lines

Due to the "finiteness" of the size of the screen, *straight lines* are *lines* given by two points, whose endpoints lie at the edge of the screen.

Rectangles

Parallel to the screen edges are *rectangles*, given by two points, the upper left and the lower right corner.

Circles

Circles are defined by their centre and their radius.

Ellipses

Only *ellipses* with *axis-parallel* semi-axes, which are defined by their centre and the lengths of their semi-axes.

Texts

Alphanumeric texts, i.e., also *numbers* (e.g., in the simplest case sequences of digits).

Image

Images are graphic files in the `ppm`-format. In order for them to be used by the electronic stylus, they naturally have to be small enough.

Their processing is possible with the routines from the `netpbm` package, e.g., their scaling with `pamscale` and the conversion between this format and the `jpeg`-format with `pnmtojpeg` or `jpegtopnm`.

Filled Figures

Rectangles, *circles*, *ellipses*, and *crossing-free convex* (under X also *arbitrary* polygons) can be *filled* in the sense that all pixels inside the figure are set to the colour (of the edge) of the figure.

9.1.2 The Operations on the eBoards

With the electronic stylus, the following operations can be performed on an eBoard—the "board images" simulated on the screen with the electronic stylus:

- Individual figures
  - *create*,
  - *modify* and *colour*,
  - *move*,
  - *delete* as well as
  - *mark* and *unmark*;
- all marked figures
  - *delete* and *unmark*,
  - *store* in another eBoard;
- all figures of an eBoard
  - *delete*,
  - *mark*;
- an eBoard
  - *move*,
  - *load* and *save*;
- another eBoard
  - *load additionally*;
- as well as all creations and deletions on an eBoard
  - *undo*.

New figures are created in the current type and the current colour, which initially have standard values and can be changed at any time with certain commands.

9.1.3 Program Start

The electronic stylus is designed according to ergonomic principles.

The program uses the screen primarily in "Fullscreen mode".

It is designed to be operated with one hand freely above the keyboard and the other hand on the mouse—apart from text input.

Under these conditions, the electronic stylus essentially uses a few keys

- Tab key ⇆,
- control key Ctrl and
- space bar,

which are close together and therefore can be used "blindly" after a short period of acclimatization, so that the focus on the eBoard is not interrupted by constant switching between screen, keyboard, and mouse.

In console operation, the control key Ctrl generally acts like the Shift key ⇑, as Ctrl is easier to reach "blindly" than ⇑.

However, the window managers of common graphical interfaces intercept certain key combinations with the Ctrlkey and the Altkey and use them to manipulate windows; a two-button mouse does not have a middle button.

Therefore, in these cases, alternative keys

- delete key Del,
- function keys F5 to F9,

which are close together and therefore can be used "blindly" after a short period of acclimatization, must be used, which somewhat contradicts the above considerations.

The operation of the electronic stylus will be explained in detail below.

9.1.4 Program Start

The electronic stylus is started with the command epen, to which the name of the eBoard can be given as a parameter. If this is the case, this name—otherwise the provisional name temp—appears in the field of the eBoard name in the top left corner of the screen; it can be edited. The corresponding files have these names with the suffix .epn.

When the input is completed with the Enter key ↵, the eBoard with this name is loaded, if there is one; otherwise, the eBoard is now empty. If no name is entered, the eBoard is named "temp".

9.1.5 Creation of New Figures

Pressing the space bar, the key A, the Enter key ↵, or the insert key Ins causes a menu to appear at the location of the mouse pointer, from which the *actual type* of figures can be selected with the mouse.

It is selected with the arrow keys ▲ and ▼ and the Home and End keys Pos1 and End and confirmed with the Enter key ↵.

The selection is cancelled with the Esckey; then the old current type remains.

New figures are drawn in this type—until another current type is selected.

Point Sequences

By pressing the left mouse button, the creation of a new figure begins at the location of the mouse pointer.

If the actual type is a *point sequence*, it will "draw"—following the mouse movement—until the mouse button is released. If the maximum possible number of points in the sequence has already been reached, the drawing will automatically end.

The faster the mouse is moved, the more noticeable the effect becomes that the individual points—recognized by the mouse—are connected by lines: The figure becomes somewhat "angular".

Lines and Line Sequences

If the current type is *line(s)* (*Lines* or *Line sequences*), the figure begins with the first mouse click.

Further clicks with the left mouse button at other locations set the next line; movements between the mouse clicks continue the last line until the next mouse click.

A line sequence is ended by a click with the right mouse button.

A *simple line* is thus created with the following sequence of commands: Move the mouse to the starting point—click with the left mouse button—move the mouse to the end point—click with the right mouse button.

Polygons

Polygons are created in a very similar way; the only difference is that from the second mouse click, the actual mouse position is automatically connected to the starting position.

If a polygon is convex (this restriction only applies to the operation of the electronic stylus in a console), it is filled in the same colour as its edge if the final click with the right mouse button is made together with the shift key ⇑.

Curves

Curves are essentially created like line sequences: Click with the left mouse button; continuation by mouse movement and further clicks with the left mouse button; the fixation is done with a click of the right mouse button.

If the maximum possible number of support points is reached during continuation the creation is automatically ended.

Unlike line sequences, the entire curve constantly adapts to the set support points during creation; it requires some practice and experience until the user "gets the hang of it".

Straight Lines
A point of a *straight line* is set with a press of the left mouse button, with the line appearing as a horizontal; the movement of the mouse leads the second point and thus the line until they are fixed by releasing the mouse button.

Rectangles
When creating *rectangles*, the procedure is similar to straight lines: A press with the left mouse button sets a corner of the rectangle; as long as the mouse is moved, the diagonally opposite corner and thus the rectangle is carried along until the rectangle is fixed by releasing the mouse button.

If the control key Ctrl is pressed while releasing the mouse button, it is filled in the same colour as its edge.

Circles
Circles are created according to the same principle: A press with the left mouse button sets the centre point; the movement of the mouse leads to the dragging of a circle through the mouse position; as soon as the mouse button is released, the circle is fixed.

Circles can be *filled* like rectangles.

Ellipses
Ellipses are created like circles; the only difference is that with the mouse movement a corner of the circumscribing rectangle is set.

Ellipses can be *filled* like rectangles.

Texts
If the actual type is a *text*, a blinking cursor appears at the position of the mouse pointer after a click with the left mouse button. The text is entered using the keyboard, and it can be comfortably modified—like with a usual editor (as described in Section 3.4.5 about the field editor of the microuniverse).

If the input is empty or is ended with a key other than the enter key ↵, the creation is aborted.

Images
If the actual type is an *image*, a blinking cursor appears—like with a text—after a click with the left mouse button, which is associated with the request to enter the name of the image.

After completing the input with the enter key ↵, the image contained in the corresponding file, whose filename is the name of the image with the appended suffix .ppm, appears with the mouse position as the top left corner, provided such a file exists and the image fits completely on the eBoard.

If the input is empty or ended with the escape key Esc, the process is aborted.

9.1.6 Modification of Figures

If the mouse pointer is on a figure, a click with the left mouse button along with the shift key ⇑ makes the points that characterize the figure visible. They can be individually "grabbed" with the right mouse button and moved; the figure adjusts accordingly. The modification is completed with a click of the left mouse button. For rectangles and circles, this is simpler; their points do not appear—they can be "grabbed" at any point on their edge.

When pressing the function key F3, a coloured strip appears, from which a colour can be selected by clicking with the left mouse button, which colours the figure under the mouse. If the mouse is moved during this, all figures that the mouse "runs over" are coloured. The selection is cancelled with a click of the left mouse button outside the strip or the escape key Esc.

If the F3 key is pressed together with the shift key ⇑, the actual colour can be selected, in which new figures will be coloured until it is changed with this procedure.

By pressing the function key F4, the background colour of the eBoard can be selected.

With the right mouse button, individual figures can be "grabbed" and moved while holding down the mouse button.

9.1.7 Deleting of Figures

If the mouse pointer is on a figure, it is deleted by pressing the delete key Del. If the mouse is moved during this, all figures that are swept over by the mouse are deleted. If the shift key ⇑ is also pressed, *all marked figures* are deleted—regardless of the mouse position.

With the backspace key ←, the last deleted figure is restored; if it was marked before, it is no longer now. Together with the shift key ⇑, all deleted figures are restored.

9.1.8 Marking Figures

A figure under the mouse pointer is marked with the function key F5. Regardless of the position of the mouse, all figures are marked when the shift key ⇑ is additionally pressed.

The marked figures flash briefly.

Correspondingly, figures are unmarked with the function key F6. When pressing the tab key ⇆, *all* marked figures flash briefly.

9.1.9 Loading and Saving

With a press of the Rollkey, all figures from another eBoard are copied into the actual eBoard. The field for the eBoard name opens, into which the name of the eBoard to be loaded is entered. The entry is completed with the Enter key ↵; when finished with the escape key Esc.

If the Rollkey is pressed together with the shift key ⇑, all marked figures are stored in another eBoard; its name is entered in the field for the eBoard name (an eBoard existing under this name will be overwritten).

9.1.10 Printing

Upon pressing the key Print, the content of the eBoard (with white background) is printed (provided CUPS is installed and a postscript-capable printer is available).

9.1.11 Brief Help

The help screen appears at the press of the function key F1, shows brief hints for operation, and disappears when pressing the escape key Esc.

9.1.12 System Architecture

The only component of the electronic pen is the "main program" epen.go with an event control in the form of an *event loop*, in which various data types from the microuniverse are used.

9.2 Construction

The most important package used from the microuniverse is that of two-dimensional figures, the specification of which we show here:

```
package fig2
import (. "µU/obj"; "µU/col"; "µU/psp")
type typ byte; const (
```

9.2 Construction

```
  Points = iota // sequence of points
  Segments // line segment[s]
  Polygon
  Curve // Bezier curve
  InfLine // given by two different points
  Rectangle // borders parallel to the screen borders
  Circle
  Ellipse // main axes parallel to the screen borders
  Text // of almost 40 characters
  Image // in ppm-format
  Ntypes
)
type Figure2 interface {

  Object
  Stringer
  Marker

// x is of typ t.
  SetTyp (t typ)

// Returns the typ of x.
  Typ() typ

// x is of the Type, that was selected interactively by the user.
  Select()

// The defining points of x are shown, iff b.
  ShowPoints (b bool)

// Returns the position of the
// - first point (of the first line), if x has a typ <= Line,
// - top left corner of x, if x is of typ Rectangle or Image,
// - middle point of x, if x is of typ Circle or Ellipse,
// - bottom left corner of first characer, if is of typ Text.
  Pos() (int, int)

// x has Position (x, y)
  SetPos (x, y int)

// Returns true, iff the point at (a, b) has a distance
// of at most t pixels from x.
  On (a, b int, t uint) bool

// x is moved by (a, b).
  Move (a, b int)

// Returns true, iff the the mouse cursor is in the interior of x
// or has a distance of not more than t pixels from its boundary.
  UnderMouse (t uint) bool

// x has the colour c.
  SetColour (c col.Colour)

// Returns the colour of x.
  Colour() col.Colour

// x is drawn at its position in its colour to the screen.
  Write()

// x is drawn at its position in its inverted colour to the screen.
  WriteInv()

// Pre: x has a typ != Image.
// x is now the figure interactively generated by the user-
  Edit()
```

```
// Pre: x has the typ Image.
// Returns true, if x is now the image interactively generated by the user.
  ImageEdited (n string) bool

// Pre: f is not empty. f != Points and f != Image.
// x is interactively changed by the user.
  Change()

// If x is a text, it has the font size f.
  SetFontsize (s fontsize.Size)

// x is printed (see package µU/psp).
  Print (psp.PostscriptPage)
}

// Returns a new empty figure with undefined typ.
func New() Figure2 { return new_() }

// Return a new empty figure with the corresponding typ.
func NewPoints (xs, ys []int, c col.Colour) Figure2 {
     return newPoints(xs,ys,c) }
func NewSegments (xs, ys []int, c col.Colour) Figure2 {
     return newSegments(xs,ys,c) }
func NewPolygon (xs, ys []int, f bool, c col.Colour) Figure2 {
     return newPolygon(xs,ys,f,c) }
func NewCurve (xs, ys []int, c col.Colour) Figure2 {
     return newCurve(xs,ys,c) }
func NewInfLine (x, y, x1, y1 int, c col.Colour) Figure2 {
     return newInfLine(x,y,x1,y1,c) }
func NewRectangle (x, y, x1, y1 int, f bool, c col.Colour) Figure2 {
     return newRectangle(x,y,x1,y1,f,c) }
func NewCircle (x, y, r int, f bool, c col.Colour) Figure2 {
     return newCircle(x,y,r,f,c) }
func NewEllipse (x, y, a, b int, f bool, c col.Colour) Figure2 {
     return newEllipse(x,y,a,b,f,c) }
func NewText (x, y int, s string, c col.Colour) Figure2 {
     return newText(x,y,s,c) }
```

Mini 10

*When you let a computer calculate,
it is to be expected,*

that it does not calculate correctly.

Abstract

Mini is a simple model of a single-address machine that introduces the basic concepts of *imperative*—especially *machine-close*—programming. It has 26 registers, 2 status flags, and 30 machine instructions for accessing the registers and status flags, for computing, and for jumping within a program.

With Mini, the *concept of state* (the values of the registers and the status flags) and the *basic algorithmic structures* (sequences, case distinctions, loops, and recursion) are introduced. The significance of this concept lies

- *theoretically* in the Turing completeness of Mini, as it is just as powerful as a register machine: Everything that can be programmed at all can, in principle, be done with Mini.
- *practically* in its clarity and comprehensibility.

Although the existing instruction set is very extensive, Mini machine programs are uncomplicated and easy to handle.

For these reasons, it was used many years ago in computer science classes at the Goethe Gymnasium in Berlin-Wilmersdorf (then still programmed in Modula-2).

10.1 System Analysis

Components of a single-address machine are

- a *processor* for executing programs,
- a *data storage* in the form of a set of *registers*,
- a *stack storage* for temporary storage of register contents, and
- a *program storage* for holding a program in the form of a sequence of *program lines*.

In addition, Mini needs a way

- to input a program, i.e., to populate the program storage with program steps and
- to populate the registers with data and output this data.

10.1.1 Processor

Mini's processor has the task

- using a *computing unit* to execute individual program steps and
- using a *control unit* to execute the program, i.e., the instructions in the sequence of program lines (details see Sect. 10.1.3).

For this purpose, it has its own *registers*:

- two *accumulator registers* ax and bx (shortly referred to as emphaccus);
- a *program counter* for controlling program execution; and
- a *status register*, in which some instructions set or clear certain *flags*, i.e., they assign the values 1 or 0 to them:
  - the *zero flag* zf and
  - the *carry flag*, which is also "misused" as an *overflow* flag.

At the beginning of the execution of a program, all registers have the value 0.

10.1.2 Data Storage

The data storage of Mini consists of 26 *registers* and the *stack memory*.

10.1 System Analysis

Registers are storage locations that can each hold a (maximum 9 digits) natural number. The contents of the registers are also referred to as their *value*.

Since the values of the registers change during the execution of a program (which is precisely the purpose of programs), the registers can also be considered as *variables*. The 26 registers have the names "a" to "z", which are used to access their values.

The *stack memory* consists—figuratively speaking—of "stacked" registers (according to the "*last in-first out*" principle); register values can be placed on top of the stack or taken from the top, with care being taken that a value can only be taken when the stack is not empty.

At the beginning of the execution of a program, the stack memory is empty.

10.1.3 Program Lines

The individual steps of a Mini machine program (hereinafter referred to as a mini program) are "line by line" in the program memory and are therefore referred to as *program lines*. They are numbered consecutively, starting at 0.

Each program line contains exactly one *instruction*, namely:

- a *storage instruction*,
- an instruction for a *computational operation*,
- a *comparison instruction*,
- a *flag instruction*,
- a *jump instruction*,
- a *stack instruction*,
- a *call instruction*, or
- a *return instruction*.

They are listed and explained in detail in Sect. 10.1.5. At the beginning of a program line, there may also be a label.

10.1.4 Execution of a Mini Program

A program is executed by a single-address computer by executing the instructions in its program lines *sequentially* (i.e., line by line), starting with the first line and then either moving to the next program line or jumping to a line other than the next.

Looking a bit more closely, this is realized as follows:

The next instruction to be executed is always the one in the particular program line whose number is in the *program counter* of the processor.

Initially, the program counter contains a 0; so it starts with the first program line. Which number *after* the execution of a program line is in the program counter, i.e., which line is processed next, depends on whether the program line contains a jump or call instruction.

A mini program is *terminated*, when it encounters a line with the *return instruction*; consequently, when creating a mini program, care must be taken to include a return instruction.

If the value of the program counter is greater than or equal to the number of program lines, the program is aborted, since the numbering starts at 0, e.g., after the last program line, if that is not a jump instruction.

A program termination can also have other causes, namely, "programming errors" in the form of unconsidered prerequisites when calling instructions, for whose execution a prerequisite is specified.

A *subprogram* consists of a sequence of program lines, the *first* of which is introduced with a label and the *last* of which consists of the return instruction. It is executed by the designated *call instruction*, with the number of the next program line being inserted into the program counter at the end of the subprogram (following the line in which the subprogram was called).

10.1.5 Instructions

Mini has a "programming language" with a few *instructions*:

- six *memory instructions* for copying register contents into/from the accumulators ax and bx:
  lda, sta, exa, ldb, stb, and exb;
- four instructions for *increasing and decreasing* the values of the accumulator and registers:
  ina, dea, inc, and dec;
- two *shift instructions* for multiplication or division of register values by or through 2:
  shl and shr;
- five instructions for executing *arithmetic operations* on the values, temporarily of the accumulators and of the registers:
  add, adc, sub, mul, and div;
- one instruction for *comparing* accumulator and register values:
  cmp;
- three *flag instructions* for manipulating the status register:
  clc, stc, and cmc;
- five *jump instructions* for "jumping" in the program, also depending on the values of the flags in the status register:
  jmp, je, jne, jc, and jnc;

10.1 System Analysis

- two *stack instructions* for temporary storage of register values:
  push and pop;
- one *call instruction* for executing a subroutine:
  call; and
- one *return instruction* for terminating a program:
  ret.

The *memory instructions* expect a register, the *jump instructions* and the *call instruction* expect a label as an argument. Instructions for *increasing* or *decreasing*, for executing *arithmetic operations* as well as *shift* and *comparison instructions* expect at most *one* and the *stack instructions* exactly one argument:

Since all instructions expect at most *one* argument, Mini is an example of a *single-address* machine.

We now provide the specifications of all instructions, where the flags are not set or cleared unless explicitly stated. After the execution of an instruction—unless otherwise stated—the value of the program counter is increased by 1, so that the next program line is processed afterwards. It is different with the return instruction, which terminates the program, and the jump instructions: The next instruction to be executed is the one in the program line that starts with *that* label, which is given as an argument to the jump instruction.

```
// ax has the value of r.
lda r

// r has the value of ax.
sta r

// The values of ax und r are changed.
exa r

// bx has the value of r.
ldb r

// r has the value of bx.
stb r

// The values of bx and r are changed,
exb

// If the value of ax was smaller than 10^9 - 1,
// it is incremented by 1, otherwise set to 0.
// zf is correspondingly set resp. cleared.
ina

// If the value of ax was greater than 0,
// it is decremented by 1, otherwise set to 10^9 - 1.
// zf is correspondingly set resp. cleared.
dea

// If the value of r was smaller than 10^9 - 1,
// it is incremented by 1, otherwise set to 0.
// zf is correspondingly set resp. cleared.
inc r
```

```
// If the value of r was greater than 0,
// it is decremented by 1, otherwise set to 10^9 - 1.
// zf is correspondingly set resp. cleared.
dec r


// The double z of the value of r is built
// and r has the value z mod 10^9.
// cf is set, iff z is greater than 10^9.
shl r


// The value of r is halved; cf is set,
// iff the value of r was previously an odd number.
shr r


// The value of ax is incremented by the value of r,
// if that is possible without overflow over 10^9 - 1;
// in this case cf is cleared. Otherwise ax has the value
// of the sum of the values of ax und r mod 10^9 and cf is set.
add r


// The value of ax is incremented by the value of r and the value of cf,
// if this is possible without overflow over 10^9 - 1.
// In this case cf is now cleared.
// Otherwise ax has now the value of the sum of the values
// of ax, r and cf mod 10^9 and cf is set.
adc r


// If the value of r is less or equal to the value of ax,
// the value of ax is decremented by this value
// and cf is cleared. Otherwise as has the value
// 10^9 - (value of r - value of ax) and cf is set.
sub r


// The product p of the values of ax and r is built:
// ax has the value p mod 10^9 and bx the value p div 10^9,
// i.e., p = value of bx * 10^9 + value of ax.
// cf is set, iff p is greater or equal to 10^9.
mul r


// Pre: r has a value > 0.
// ax has the value q div r and bx the value q mod r.
div r


// The programcounter shows the number of
// the first program line with the label M.
// If there is not program line with this label,
// the program is aborted.
jmp M


// Pre: There is a program line that starts with M,
// and this program line is not the last one.
// If zf is set, the program counter shows the number
// of the first program line with the label M,
// otherwise the number of the following program line.
je M


// Pre: s. je.
// If zf is cleared, the program counter shows the number
// of the first program line with the label M,
// otherwise that one of the following program line.
jne M
```

10.1 System Analysis

```
// Pre: s. je.
// If cf is set, the program counter shows the number
// of the first program line with the label M,
// otherwise that one of the following program line.
jc M

// Pre: s. je.
// If cf is cleared, the program counter shows the number
// of the first program line with the label M,
// otherwise that one of the following program line.
jnc M

// If the value of r is equal to the value of ax,
// is zf is set. Otherwise it is cleared and cf is set,
// iff the value of r is smaller than the value of ax.
cmp r

// cf is cleared, i.e., has the value 0.
clc M

// cf is set, i.e., has the value 1.
stc M

// cf is set, i.e., has the value 1.
// cf is complemented, i.e. is set,
// if cf was previously cleared, and vice versa.
cmc M

// cf is set, i.e., has the value 1.
// The value of r is pushed on the stack.
push r

// cf is set, i.e., has the value 1.
// If the stack did not contain a value, the program
// is aborted with a corresponding error report.
// Otherwise, r now has the top value of the stack
// and this value is removed from the stack.
pop

// cf is set, i.e., has the value 1.
// The program counter contains the number
// of the first program line with the label M.
// The next return-instruction implies that
// the program counter contains the number
// the follows the line the call-instruction.
call M

// cf is set, i.e., has the value 1.
// The program resp. the subprogram has ended.
ret
```

For the name r of the register used in these specifications, the number of any register can be inserted in the instructions; the same applies to the used label M. In this sense, these lines are to be understood as *templates* for instructions.

10.1.6 Example

The following program calculates the factorial of the value of b and writes it into the register a, if a initially contains the value 1:

```
      lda a
A     mul b
      sta a
      dec b
      jne A
      ret
```

However, this example is only correct if the value of b is less than or equal to 12, because otherwise the result is $\geq 10i^9$.

A bit further, the following mini-program works:

```
A:
   lda a
   mul b
   sta a
   stb c
   lda d
   mul b
   add c
   sta d
   dec b
   jne A
   ret
```

The result mod 10^9 (i.e., the 9 low places) is in the register a after the execution of the program, the result div 10^9 (i.e., the up to 9 high places) in the register d.

The readers should convince themselves by recalculating this mini-program is correct up to the start value 19 of b.

10.1.7 The Objects of the System

The following *objects* can be derived from the system analysis:

- *program* as a sequence of program lines,
- *program lines*, and
- *register*.

The implementation of the single-address machine Mini therefore consists of the package mini with the subpackages:

- mini/prog,
- mini/line, and
- mini/reg.

10.1.8 Component Hierarchy

The dependencies of the packages are shown in Fig. 10.1, where the lower package is used (imported) by the one above:

Fig. 10.1 System architecture of Mini

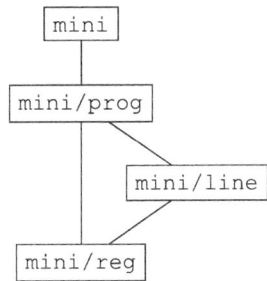

At lower levels, many other packages from the microuniverse are needed, e.g., *input/output fields* (μU), *stack memory* (μU), and *persistent sequences* (μU).

10.2 User Manual

Mini is a simulation program for executing mini programs. For the identifiers in mini programs, whose execution is to be simulated by Mini, the following conventions must be observed:

- *variables*, i.e., names of registers, are designated with a lowercase letter (from a to z) and
- *labels* with a capital letter (from A to Z).

The operation of the program is extremely simple.

In addition to the letter, number, and character keys for entering text, some special keys are needed for correcting entries and controlling the program sequence.

The following keys are used for input correction:

- the backspace key ← and the delete key Del for deleting individual characters, in combination with the shift key ⇑ to delete the input field;
- the arrow keys ◀ and ▶ to the left and right;
- the home key Pos1 and the end key End for moving in the text; and
- with the insert key Ins you can switch between insert and overwrite modes, with the current mode being recognizable by the different cursor shape: an underscore in insert mode and a rectangular block in overwrite mode.

The program flow is controlled with

- the enter key ↵, the escape key Esc, the backspace key ←;
- the arrow ▲ and ▼ and the page keys, Page↑ and Page↓ up and down, as well as
- the tab key ⇆,

occasionally in combination with the shift key ⇑. Error messages are acknowledged with the backspace key ←.

10.2.1 Instructions for Working with Mini

- Call Mini by entering `mini`, where the name of the mini program to be executed (without the extension `.mini`) is given as a parameter, e.g., `mini test`;
- if necessary, edit the mini program;
- exit the edit mode with the Esc key Esc;
- enter the initial values of the used registers;
- step through the mini program with the enter key ↵ (the register contents are continuously displayed) and—if desired—exit the step mode with Esc and let the mini program run to the end;
- if desired, abort the program execution of Mini with the combination Ctrl + C; and
- after execution of the program (which is acknowledged with the message `program executed`, exit Mini with Esc.

A parameter given with the program call `mini` is the name of the mini program, otherwise it gets the name `prog`. The associated *files* have the extension `.mini` and are stored in the subdirectory `.mini` of the home directory `$HOME`.

10.3 Construction

Here are the specifications of the three packages used by `mini`:
```
package prog
type Program interface {
  Empty() bool
  GetLines()
  Parse() (string, uint)
  Write()
  Edit()
  Run()
}

func New() Program { return new_() }
```

```
package line
import . "µU/obj"

const EmptyLabel = byte(' ')

type Instruction byte; const (
NOP = Instruction (iota)
LDA; STA; LDB; STB; EXA; EXB // Argument: Register
INA; DEA
// no argument
INC; DEC
// argument: register
SHL; SHR
// argument: register
ADD; ADC; SUB; MUL; DIV
// argument: register
CMP
// argument: register
JMP; JE; JNE; JC; JNC
// argument: label
PUSH; POP
// argument: register
CLC; STC; CMC
// no argument
CALL; RET
// no Argument
nInstructions
)

type Line interface { // lines of mini programs, consisting of a label,
// an instruction, a register and a target label.

  Clearer
  Equaler
  Stringer
  Write (l, c uint)
  Edit (l, c uint)

// Returns (M, true), iff x starts with label M;
// returns otherwise (EmptyLabel, false).
  Marked() (byte, bool)

// Returns true, iff x contains the instruction CALL.
  IsCall() bool

// Returns true, iff x contains the instuction RET.
  IsRet() bool

// The instruction in x is executed.
  Run() byte
}

func New() Line { return new_() }
// The state of the processor is written to the screen,
// starting at position (line l, column c).
func WriteStatus (l, c uint) { writeStatus(l,c) }
```

10.4 Exercises

```
package reg
import . "µU/obj"
```

```
const (N = 9; M = 1e9; R = 26) // R = number of registers

func New() Register { return new_() }

type Register interface { // register with a small character
// "a", ..., "z" as mames and a natural number < N as value.
// All R registers are managed.

  Clearer
  Stringer
  Valuator

  Write (l, c uint)
  Edit (l, c uint)
}

func WriteAll (l, c uint) { writeAll(l,c) }
func EditAll (l, c uint) { editAll(l,c) }
```

and here is the event loop of the main program:

```
package main
import ("µU/mode"; "µU/scr"; "µU/errh"; "mini/prog")

func main () {
  scr.New (0, 0, mode.VGA); defer scr.Fin()
  program := prog.New()
  program.GetLines()
  fail, n := program.Parse()
  if fail == "" {
    program.Write()
    program.Edit()
    program.Run()
  } else {
    errh.Error (fail + " <- faulty program line nr.", n + 1)
  }
}
```

Develop mini programs for calculating

- the power of two numbers (e.g., 2²⁰),
- the sum and product of two or more numbers,
- the quotient of two numbers,
- the minimum/maximum of two or more numbers,
- the GCD and LCM of two numbers,
- the sum and product of two fractions, and
- of binomial coefficients and Fibonacci numbers.

Books   11

> *That which you seek,*
> *you always find in the place,*
> *where you look last.*
>
> One of the laws of
> Edward A. Murphy jr.

Abstract

This program is a system for managing a book inventory, constructed as a simple application of persistent index sets. Edward A. Murphy jr.

The program presented here is a somewhat simplified version of a teaching project from the teacher training in computer science at the Free University of Berlin. It was about creating a system for managing any collection (see Chapter *Inferno*).

Here we limit ourselves to the special case of a collection of books.

11.1 System Analysis

The following data should be recorded for each book:

- *area* (prose, classic, Novel, …),
- *author*,
- *co-author*,
- *number (for series)*,
- *title*, and
- *location*.

Examples:

- theatre,
- Dürrenmatt, Friedrich,
- _,
- _,
- The Physicists,
- 1st shelf in the library (in this case there is no co-author and no series number).

or

- Italian crime novel,
- Fruttero, Carlo,
- Lucentini, Franco,
- 4,
- The Secret of the Pineta,
- small bookcase.

Three orders are provided:

- area,
- authors, and
- titles.

11.2 System Architecture

11.3 The Objects of the System

Thus, the system has the following objects:

- an enumeration type *Areas* for the first component mentioned in the system analysis;
- *strings* for the 2nd, 3rd, 5th, and 6th component;
- *natural numbers* for the 4th component;
- the *compound*, which combines these components; and
- the *program* books.

The corresponding packages are field, text, bn, book, and books.

Fig. 11.1 System architecture of the management of the book inventory

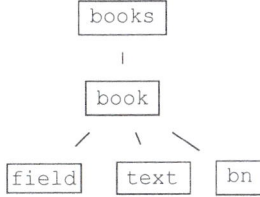

11.4 Component Hierarchy

The dependencies of these packages on each other are shown in Fig. 11.1.

11.5 User Manual

With ▲ and ▼ you can scroll backward or forward through the inventory, and with Pos1 and End you can go to the first or last entry. The actual entry can be changed after entering ↵.

Pressing Ins causes an empty entry to appear on the screen, whose input fields can be filled out; this entry is then inserted into the inventory.

With Del, the actual entry is removed after a security query, with F3 the order is changed and with Esc the program is terminated and the data inventory is secured.

Figure 11.2 shows the screen mask.

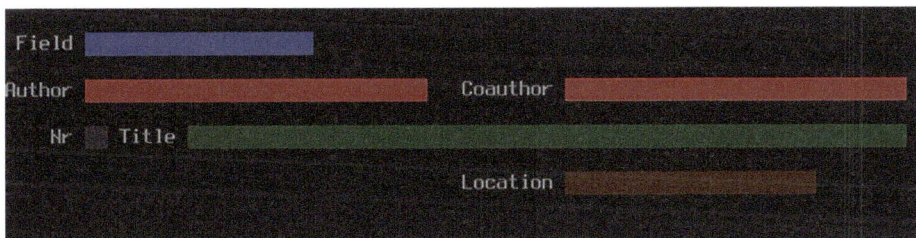

Fig. 11.2 The screen mask

11.6 Construction

Here is the specification of the data types used:

11.6.1 Areas

Unfortunately, Go does not have an enumeration type; therefore, we make do with a sequence of constants that are defined only in the implementation, but are not visible in the following specification:

```
package field
import . "µU/obj"

type Field interface {

  Object
  Editor
  Stringer
}

func New() Field { return new () }
```

Here is the representation and the constructor:

```
const (Undef = iota; Prosa; Klassik; Roman; ...)

var tx = [nFields]string {"prose"
                          "classics"
                          "roman"
                          ...
                          }
type field struct {
                int "order number of the constant"
                text.Text
}

func new_() Field {
  x := new(field)
  x.int = Undef
  x.Text = text.New (18)
  x.Text.Defined (tx[Undef])
  return x
}
```

Anyone who wants to use the program books should adapt the implementation to their personal needs.

11.6.2 Natural Numbers

We consider *natural numbers* with a fixed maximum number of digits:

```
package bn
import (. "µU/obj"; "µU/col")

const M = 20 // 1<<64 - 1 = 18446744073709551615 has 20 digits
```

11.6 Construction

```
type Natural interface { // natural numbers < 2^64 - 1.

  Object
  col.Colourer
  EditorGr
  Stringer
  Valuator
  Printer

// Returns the width of x given by New.
  Width() uint

// Pre: s contains only the digits 0 and 1.
// x is the natural number with the binary represenation s.
  Decimal (s string)

// Returns the binary representation of x.
  Dual() string
}

// Pre: 0 < n <= M.
// Returns a new Natural with value 0 for numbers with at most n digits.
func New (n uint) Natural { return new_(n) }
```

Here is the representation and the constructor:

```
const invalid = uint(1<<64 - 1)

type natural struct {
                    uint
                 wd uint
            f,  b col.Colour
                 font.Font
}

func new_(n uint) Natural {
  if n == 0 || n > M { ker.PrePanic() }
  x := new(natural)
  x.uint = invalid
  x.wd = n
  x.f, x.b = col.StartCols()
  return x
}
```

11.6.3 Strings

Texts are short strings that fit on one line of a screen. Here is the specification of the corresponding package:

```
package text
import (. "µU/obj"; "µU/col")

type Text interface { // strings of fixed length

  Editor
  col.Colourer
  Stringer
  Texer
  Printer

// Specs see str/def.go.
  Equiv (Y Text) bool
```

```
  Transparence (t bool)
  SetFontsize (s fontsize.Size)
  SetFont (f font.Font) // only to print

  Sub (Y Text) bool
  Sub0 (Y Text) bool
  EquivSub (Y Text) (uint, bool)
  Len() uint
  ProperLen() uint
  Byte (n uint) byte
  Pos (b byte) (uint, bool)
  Replace1 (p uint, b byte)

// starting with position p in x, n bytes are removed;
// tail filled with spaces up to the original length
  Rem (p, n uint)

  IsUpper0() bool
  ToUpper()
  ToLower()
  ToUpper0()
  ToLower0()

  Split() []Text

  WriteGr (x, y int)
  EditGr (x, y int)
}

// Returns a new empty text of length n.
func New (n uint) Text { return new_(n) }

// Returns a new text of length len(s) with the content s.
func Def (s string) Text { return def(s) }
```

and here is their representation and the constructor:

```
type text struct {
                uint "length of string"
                string
         cF, cB col.Colour
                font.Font
                font.Size
}

func new_(n uint) Text {
  x := new (text)
  x.uint = n
  x.string = str.New (n)
  x.f, x.b = col.StartCols()
  x.Size = fontsize.Normal
  x.Font = font.Roman
  return x
}
```

11.6.4 Book

Here is the specification of the type book:

```
package book
import . "µU/obj"
```

11.6 Construction

```
type Book interface {

  Indexer
  Rotator

// Pre: y is of type Book.
// Returns true, iff x is a part of y.
Sub (y Any) bool
}

func new_(n uint) Text {
  if n == 0 { return nil }
  x := new (text)
  x.uint = n
  x.string = str.New (n)
  x.cF, x.cB = col.StartCols()
  x.Font = font.Roman
  x.Size = font.Normal
  return x
}

  func New() Book { return new_() }
```

11.6.5 Books

Its representation is the composite of the presented components:

```
type book struct {
                field.Field
author, coauthor text.Text
                bn.Natural
 title, location text.Text
}

func Id (a Any) Any {return a}
```

The identity used as the index function

```
func new_() Book {
  x := new (book)
  x.Field = field.New()
  x.author = text.New (len0)
  x.coauthor = text.New (len0)
  x.Natural = bn.New (2)
  x.title = text.New (len1)
  x.location = text.New (len0)
  return x
}
```

11.6.6 The Program for Managing the Book Inventory

For the control of the program, the function Operate from the package μU is used, whose implementation is algorithmically uninteresting which is why we do not go into it here. This makes it very short:

```
package main
import (. "µU/collop"; "µU/env"; "µU/scr"; "µU/files"
        "µU/piset"; "µU/collop"; "µU/book")

func sub (x, y Indexer) bool {
  return x.(book.Book).Sub (y.(book.Book))
}

func main() {
  scr.NewWH (0, 0, 80 * 8, 10 * 16); defer scr.Fin()
  files.Cds()
  b := book.New()
  p := piset.New (b)
  p.Name (env.Call())
  collop.Operate (p, b, sub)
}
```

Inferno   12

> *All that is abstract,*
> *is brought closer to human understanding through application,*
> *and thus human understanding*
> *reaches abstraction through action and observation.*
>
> Johann Wolfgang von Goethe
> From Wilhelm Meisters Wanderjahre

Abstract

The Inferno is, in a sense, an abstraction of the book project. It serves to manage arbitrarily configurable data sets. Data sets can be found based on those attributes that were determined as an index during the construction of the Inferno program, i.e., intended for this purpose. The operation of the system is very simple. It can be used for many purposes, e.g., for an address directory or for managing a collection of sound carriers.

As part of teacher training and further education, I used to ask people from the industry to give a lecture on their work on certain topics. Once, the lecture of an IBM employee was combined with a visit to his workplace in the IBM building at Ernst-Reuter-Platz in Berlin. He introduced us to a program that IBM developers had been working on for several months.

In this chapter, I present the work of the colleagues from this further education course, in which we "reconstructed" this program.

12.1 System Analysis

The "Books" project from the previous chapter is to be generalized in such a way that data sets with almost arbitrary components can be recorded, browsed, searched, and found, as well as deleted.

The basic components of the system are

- its masks,
- its data sets,
- their structure, and
- components.

▶ In this chapter, we call the data sets "molecules" and the components they are made up of, "atoms".

In the following, we explain the four basic components.

12.1.1 Masks

By "masks" we understand the invariant parts of the screen that, so to speak, name the atoms.
In the Books project, these are the following components of the screen window:

- `area`,
- `author`,
- `co-author`,
- `number`,
- `title`, and
- `location`.

12.1.2 Molecules

Molecules are the contents of the system's data sets.

▶ A molecule must consist of at least two atoms.

Exactly one molecule is always displayed in the screen window.

12.1.3 Structure of the Molecules

The structure of the molecules consists of the sequence of the structures of its atoms. For each atom, this includes the following information:

- its type,
- the position on the screen window (row and column),
- its (column) width,

- its foreground and background colour, and
- the indication whether it is an index or not.

12.1.4 Atoms

The atoms can have the following types:

- strings (of a maximum of 64 characters),
- natural numbers (with a maximum of 10 digits),
- real numbers (with a maximum of 20 digits),
- calendar dates (in the form "dd.mm.yyyy"),
- times (in the form "hh.mm"),
- amounts of money (up to 10 million),
- telephone numbers (with a maximum of 16 digits incl. spaces),
- names of countries, and
- enumeration types: sequences of strings of a maximum length of 20, from which one is always selected.

In the Books project, only the first two types are used.

12.2 System Architecture

12.2.1 The Objects of the System

The *system architecture* initially provides as *objects* the

- the *molecules* and
- the *atoms*.

In addition, there are the objects that can be types of atoms.

For the construction of the program `inferno`, in addition to the packages `pseq` for persistent sequences and `set` for ordered sets, abstract data types for the molecules, their structure, and their atoms are needed. Together with the packages used by the package `atom`, we therefore have the packages

- `mol` for molecules,
- `stru` for their structure,
- `atom` for atoms,
- `text` for texts,

- `bn` for natural numbers,
- `br` for real numbers,
- `day` for calendar dates,
- `clk` for times,
- `euro` for amounts of money,
- `phone` for phone numbers,
- `cntry` for countries, and
- `enum` for enumeration types.

Of course, many more subpackages of the microuniverse μU are needed, which we will not go into here.

12.2.2 Component Hierarchy

The dependencies of the packages are shown in Fig. 12.1, where the lower package is used by the one above it.

12.2.3 The Objects of the System

They result directly from the previous considerations:

- masks,
- molecules,
- molecule structure, and
- atoms.

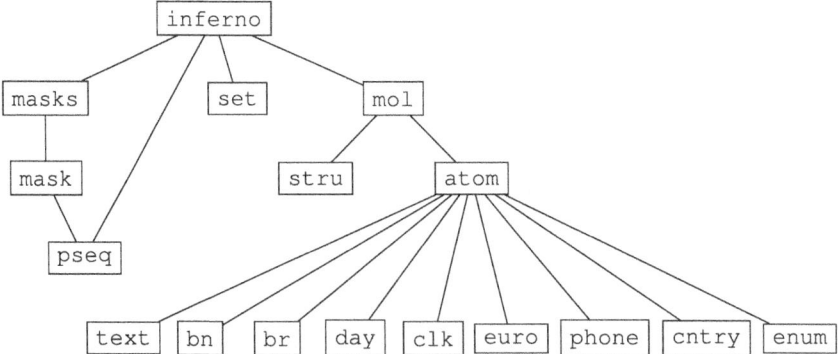

Fig. 12.1 System architecture of Inferno

12.3 User Manual

The user manual consists of two parts:

- the construction of an Inferno program and
- the operation of the system (after its construction).

12.3.1 Construction of an Inferno Program

The construction of an inferno program consists of three steps:

- the determination of the window size,
- the creation of its masks, and
- the construction of the structure of its molecules.

12.3.1.1 Setting the Window Size
To create a new Inferno program, the appearance of the window must first be designed.

The design specifies the masks with their positions and all atoms of the molecules with their type, their positions in the window, and their length.

As an example, we choose an address directory. Its molecules consist in Germany of the following atoms (for other countries the format of Address should probably be changed):

- first and last name,
- date of birth,
- address (street/no., zip code, city), and
- phone number.

Figure 12.2 shows the window of our example.

In this example, the window has the following masks:

- *name* at position (0, 0),
- *first name* at position (0, 33),

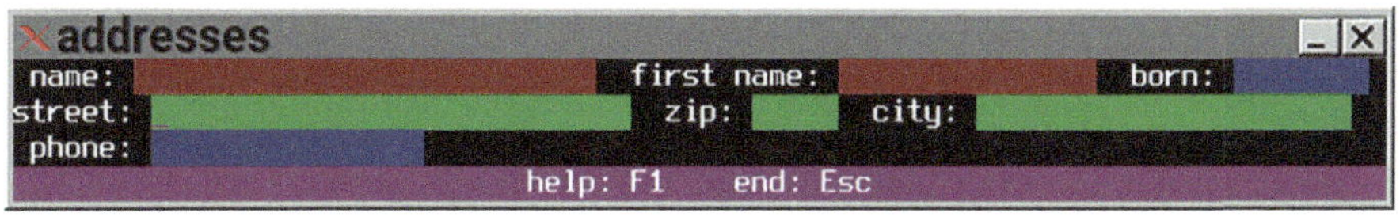

Fig. 12.2 Window of the example

- *born* at position (0, 57),
- *street/No.* at position (1, 0),
- *zip* code at position (1, 37),
- *city* at position (1, 51), and
- *phone* at position (2, 1).

The molecules have seven atoms, which can take the following values:

1. strings of up to 27 characters,
2. strings (up to 15 characters),
3. calendar dates (8 characters),
4. strings (up to 27 characters).
5. natural numbers (5 digits),
6. strings (up to 19 characters), and
7. phone numbers (up to 16 digits/spaces).

Since a line must be provided at the bottom of the window for hints and error messages, the number of lines must be chosen one larger than necessary for the masks and molecules. In our example, the window must therefore have 4 lines and 71 columns.

Every Inferno program must have a name. The name must not contain any spaces.

The window size is created by the call "`inferno name h w`", where `name` is the name of the program, `h` is the number of lines, and `w` is the number of columns of the window for the inferno program.

Let's assume our example program is to be named "addresses". Then the construction of the masks is called with the command "`inferno addresses 4 71`", which creates a window with 4 lines and 71 columns.

12.3.1.2 Generation of Masks

The second step of the construction consists of generating the masks, for which the design is needed.

After the aforementioned call, a window of the chosen size becomes visible, in which the mouse pointer can be seen. The hint "`Edit masks`" appears. The actual position is continuously displayed at the bottom left. We move the mouse to the desired starting position of the first mask (in our example (0, 0)) and then click with the left mouse button. Now the first mask "`Name`" can be entered (from this position), which is completed with the enter key ↵.

We generate the other masks accordingly by clicking on their starting positions.

In our example, we generate the second mask by clicking on the position (1, 0) and then the input "first name".

When all masks have been created, we finish the construction of the masks with the escape key. This generates two files:

- one named "`addresses.h.dat`" for storing the window size and
- one named "`addresses.msk`", in which these masks are stored for use in our Inferno program.

12.3.1.3 Construction of the Structure of the Molecules

The second step in constructing an Inferno program is to define the structure of its molecules. They consist for each of its atoms of

- its type (see previous section),
- its position,
- its length, and
- the indication whether it is an index or not.

After creating the masks—or if they were aborted with the escape key, after calling `inferno name`—the masks appear with the hint "`Molecule construction`".

We proceed as with the mask generation: We move the mouse to the desired starting position of the first atom (in our example the name). A click with the left mouse button then leads us to a pop-up menu, in which we can move between the possible types for the atom using the arrow keys ▼ and ▲ and the keys Pos1 and End and select the desired type with the enter key ↵.

If the type "`string`" was selected, the maximum number of its characters must be entered. If the type "`natural number`" or "`real number`" was selected, the maximum number of digits or the digits before the decimal point must be entered.

Then the hint "`(Shift-)Enter: (no) index`" appears, i.e., if the enter key is pressed afterwards, this atom becomes an index. If this is not desired, the enter key must be combined with the shift key.

▶ At least one atom must be an index!

The role of the indices will be explained in the next section.

If the enumeration type Enum was selected, the hint "`Enter strings`" appears. Then the strings from which the enumeration type consists must be entered (end with the escape key).

Then the note "`Select background colour`" appears; we select it as in a pop-up menu.

We repeat these steps until the structures of all atoms are defined and finish the construction with the escape key.

The result of this step is—in our example—a file named "addresses.s.dat", in which the data constructed in this step are stored.

If desired, molecules can now be entered, otherwise the construction is aborted with the escape key.

The molecules are in the file `name.seq` (`name` = name of the inferno program).

12.3.2 System Operation

When the fully constructed inferno program is called up by entering "`inferno name`" (where `name` is its name), the first molecule appears in the window, if there is one; otherwise, only its masks and empty input fields in the constructed background colours are visible.

The pressing of one of the following command keys is expected, each of which is indicated what it does:

- Enter key ↵: change molecule,
- Esc: exit program,
- ▲/▼: scroll to previous/next molecule,
- Pos1/End: scroll to first/next molecule,
- Ins: insert new molecule,
- Del: remove displayed molecule (with safety query),
- F2: search for molecule,
- F3: advance current index, and
- Print: print displayed molecule.

When entering an atom of the type "`enumeration type`", a pop-up menu appears from the strings defined in the construction. We select the desired one (the one with the previously defined background colour) using the Enter key. The selection and thus an input can be cancelled with the escape key.

A record can be searched for by entering the value to be searched for in the field of an atom that is an index after entering F2. If there are molecules in which this value occurs, the first one is displayed. If there are several, you can scroll up/down between them using the arrow keys. For enumeration types, only one of the defined strings can be searched for.

If there are several atoms that are an index, F3 will make the next one in the circle of indices the actual index. The order of output of the molecules when scrolling through them is determined by the actual index.

In our example, the atoms "`Name`" and "`Location`" are indices. Initially, the name is the actual index, so the order when scrolling through the molecules is determined by the alphabetical order of the names and after entering F3 by that of the locations.

12.3 User Manual

If "ei" is entered for the name after entering F2, "Meier" and "Einstein" are found, for example, if they exist.

After entering F3, the addresses are scrolled through in alphabetical order of the locations.

12.3.3 Construction

The packages mentioned in the section on system architecture are components of the microuniverse μU We will introduce them in the following.

12.3.4 Molecules

The specification of the data type Molecule in the molecule package mol is as follows:

```
package mol
import . "µU/obj"

const Suffix = ".s.dat"
type Molecule interface { // structs with atoms as components.
                          // Any molecule has at least one order.
                          // One of the orders is always
                          // the actual one.
  Object

  DefineName (n string)
// Pre for Edit: DefineName has to be called immediately before.
  Editor
  Print()
  NumAtoms() uint
  Indexer
  Rotator

  Sub (Y any) bool
  Construct (n string)
}

// Returns a new empty moledule.
func New() Molecule { return new_() }

// Returns the molecule that was built by the call of Construct.
func Constructed (n string) Molecule { return constructed(n) }
```

The data type Molecule for the molecules defined in the package mol is the following:

```
type molecule struct {
                uint // length of a
              a []atom.Atom
                }
```

We now show the representation of the molecules and some essential functions for the implementation of the main program inferno.go.

```
package mol
import (. "µU/obj"; "µU/kbd"; "µU/col"; "µU/scr"
        "µU/errh"; "µU/pseq"; "µU/atom"; "µU/stru")

type molecule struct {
```

```
                    uint  // length of a
        a []atom.Atom
                    }
var (
  index []uint
  nIndices uint
  actIndex uint
  file pseq.PersistentSequence
)

func new_() Molecule {
  x := new(molecule)
  x.a = make([]atom.Atom, 0)
  return x
}

func (x *molecule) Less (Y any) bool {
  y := x.imp (Y)
  if x.a[actIndex] != y.a[actIndex] {
    return x.a[actIndex].Less (y.a[actIndex])
  }
  for i := uint(0); i < nIndices; i++ {
    if i != actIndex {
      if x.a[i] != y.a[i] {
        return x.a[i].Less (y.a[i])
      }
    }
  }
  return false
}

func (x *molecule) defineIndices() {
/*/ example for len(x.a) = 6:
  If x.a[i].IsIndex() for the numbers marked by "*",

      0   1   2   3   4   5
    |---|---|---|---|---|---|
    |   | * | * |   | * |   |
    |---|---|---|---|---|---|

  then nIndices = 3, index[0] = 1, index[1] = 2 and index [2] = 4.
/*/
  nIndices = 0
  for i := uint(0); i < x.uint; i++ {
    if x.a[i].IsIndex() {
      index = append (index, uint(0))
      index[nIndices] = i
      nIndices++
    }
  }
  actIndex = index[0]
}

func (x *molecule) Construct (name string) {
  errh.Hint ("construction of molecules")
  i := uint(0)
  loop:
  for {
    x.Write (0, 0)
    cmd, _ := kbd.Command()
    scr.MousePointer (true)
    l, c := scr.MousePos()
    switch cmd {
    case kbd.Esc:
      if nIndices == 0 {
        errh.Error0 ("no index !")
      } else {
```

12.3 User Manual

```
          break loop
        }
      case kbd.Here:
        x.uint++
        a := atom.New()
        x.a = append (x.a, a)
        x.a[i] = a
        x.a[i].Place (l, c)
        x.a[i].Select()
        if x.a[i].Typ() == atom.Enum {
          x.a[i].EnumSet (l, c, name, i)
        }
        x.a[i].EditIndex()
        x.a[i].Index (x.a[i].IsIndex())
        if x.a[i].IsIndex() { nIndices++ }
        errh.Hint ("select backgroundcolour")
        x.a[i].SelectColB()
        errh.Hint ("construction of molecules")
        i++
      case kbd.Go:
        l0 := scr.NLines() - 1
        scr.Colours (col.FlashWhite(), col.Black())
        scr.Write ("       ", l0, 0)
        scr.WriteNat (l, l0, 0)
        scr.WriteNat (c, l0, 4)
      }
    }
    errh.DelHint()
    x.defineIndices()
// store the structure of x
    file = pseq.New (stru.New())
    file.Name (name + Suffix)
    for i := uint(0); i < x.uint; i++ {
      s := stru.New()
      w := x.a[i].Width()
      s.Define (x.a[i].Typ(), w)
      l, c := x.a[i].Pos()
      s.Place (l, c)
      f, b := x.a[i].Cols()
      s.Colours (f, b)
      s.Index (x.a[i].IsIndex())
      file.Seek (i)
      file.Put (s)
    }
  }

// Returns the molecule constructed from the stored structure
func constructed (name string) Molecule {
  file = pseq.New (stru.New())
  filename := name + Suffix
  file.Name (filename)
  m := new_().(*molecule)
  num := file.Num()
  m.uint = num
  m.a = make([]atom.Atom, num)
  for i := uint(0); i < num; i++ {
    file.Seek (i)
    s := file.Get().(stru.Structure)
    m.a[i] = atom.New()
    m.a[i].Define (s.Typ(), s.Width())
    if m.a[i].Typ() == atom.Enum {
      m.a[i].EnumGet (name, i)
    }
    l, c := s.Pos()
    m.a[i].Place (l, c)
    f, b := s.Cols()
    m.a[i].Colours (f, b)
```

```
    m.a[i].Index (s.IsIndex())
  }
  m.defineIndices()
  return m
}

func (x *molecule) NumAtoms() uint {
  return uint(len(x.a))
}

func (x *molecule) Index() Func {
  return func (a any) any {
    x, ok := a.(*molecule)
    if ! ok { TypeNotEqPanic (x, a) }
    return actIndex
  }
}

func (x *molecule) Rotate() {
  actIndex = (actIndex + 1) \% nIndices
}
```

12.3.5 Structure

Here is the specification of the type Structure:

```
package stru
import (. "µU/obj" "µU/col")

type Structure interface {
// Sextuples of an atom-typ, a position on the screen,
// a fore- and a background colour and a boolean value
// indicating the structure is an index.

  Object

  Colours (f, b col.Colour)
  Cols() (col.Colour, col.Colour)
  Define (t int, n uint)
  Typ() int
  Index (b bool)
  IsIndex () bool
  Place (l, c uint)
  Pos() (uint, uint)
  Width() (uint)
}

func New() Structure { return new_() }
```

Its representation is defined in the implementation as follows:

```
package stru
import (. "µU/obj"; "µU/col"; "µU/scr")

type structure struct {
                   int // typ - see µU/atom
          l, c, w uint
              f, b col.Colour
                  bool // isIndex
                  }

func new_() Structure {
```

```
    x := new(structure)
    x.int = 0 // typ atom.String
    x.f, x.b = col.FlashWhite(), col.Black()
    return x
}
```

12.3.6 Atoms

The data type for the atoms of the molecules is the type Atom, which is specified in the package atom as follows:

```
package atom
import (. "µU/obj"; "µU/col")

const (
  String = iota; Natural; Real; Calendarday; Clocktime;
         Euro; PhoneNumber; Country; Enum; Ntypes
)
type Atom interface {

  Object
  col.Colourer
  Write()
  Edit (n string, i uint)
  EditIndex()
  Print (l, c uint)
  Place (l, c uint)
  Pos() (uint, uint)
  Width() uint
  PosLess (Y any) bool
  String() string
  Index (b bool)
  IsIndex() bool

// Pre: If x has type Enum, x.EnumSet must have been called before.
// x is the atom interactively selected by the user.
  Select()

// Pre: t < NTypes
// x has the type t and width n.
  Define (t int, n uint)

// Returns the type of x.
  Typ() int

// If x has the type String, true is returned, iff x is a substring of Y.
// Returns otherwise true, iff x.Eq (Y).
  Sub (Y any) bool

  SelectColF()
  SelectColB()

  EnumSet (l, c uint, n string, i uint)
  EnumGet (n string, i uint)
}

// Returns a new atom of type Char.
func New() Atom { return new_() }
```

We show here its representation and as an example for the implementations of the functions the one of Copy.

```
package atom
import ("µU/ker"; . "µU/obj"; "µU/kbd"; "µU/col"; "µU/scr"; "µU/str"
  "µU/box"; "µU/errh"; "µU/sel"; "µU/N"; "µU/bn"; "µU/br"; "µU/text"
  "µU/day"; "µU/clk"; "µU/euro"; "µU/phone"; "µU/cntry"
  "µU/atom/internal")

const M = 64 // maximal string length
type atom struct {
                int // typ
                text.Text
                bn.Natural
                br.Real
                day.Calendarday
                clk.Clocktime
                euro.Euro
                phone.PhoneNumber
                cntry.Country
                enum.Enum
        l, c, w uint
           f, b col.Colour
                bool // is index
                }
var (
  w = []string {"string             ",
                "natural number     ",
                "real number        ",
                "date               ",
                "time               ",
                "money amount       ",
                "phone number       ",
                "country            ",
                "enumeration type"}
  wlen = uint(len(w[0]))
  bx = box.New()
)

func new_() Atom {
  x := new (atom)
  x.int = String
  x.f, x.b = col.FlashWhite(), col.Blue()
  x.w = 1
  return x
}

func (x *atom) Copy (Y any) {
  y := x.imp (Y)
  x.int = y.int
  x.l, x.c, x.w = y.l, y.c, y.w
  x.f.Copy (y.f)
  x.b.Copy (y.b)
  x.bool = y.bool
  switch y.int {
  case String:
    x.Text = text.New (y.Text.Len())
    x.Text.Copy (y.Text)
  case Natural:
    x.Natural = bn.New (y.Natural.Width())
    x.Natural.Copy (y.Natural)
  case Real:
    x.Real = br.New (y.Real.Width() - 4)
    x.Real.Copy (y.Natural)
  case Calendarday:
    x.Calendarday = day.New()
    x.Calendarday.Copy (y.Calendarday)
  case Clocktime:
    x.Clocktime = clk.New()
    x.Clocktime.Copy (y.Clocktime)
```

12.3 User Manual

```
    case Euro:
      x.Euro.Copy (y.Euro)
      x.Euro.Copy (y.Euro)
    case PhoneNumber:
      x.PhoneNumber = phone.New()
      x.PhoneNumber.Copy (y.PhoneNumber)
    case Country:
      x.Country = cntry.New()
      x.Country.Copy (y.Country)
    case Enum:
      x.Enum = enum.New (x.w)
      x.Enum.Copy (y.Enum)
  }
}
```

And finally, an excerpt from the source code of inferno.go:

```
package main
import ("µU/env"; "µU/ker; . "µU/obj"; "µU/kbd"; "µU/scr"; "µU/str"
        "µU/errh"; "µU/files"; "µU/pseq"; "µU/set"; "µU/masks"; "µU/mol")

func sub (x, y Rotator) bool {
  return x.(mol.Molecule).Sub (y.(mol.Molecule))
}

func main() {
  files.Cds()
  ms := masks.New()
  name := env.Arg(1)
  str.OffSpc (&name)
  if name == "" { ker.Panic ("...") }
  ms.Name (name)
  if pseq.Length (name + mol.Suffix) > 0 && env.NArgs() > 1 { ker.Panic ("...") }
  h_file := pseq.New (uint(0))
  h_file.Name (name + ".h.dat")
  var w, h uint
  if ms.Empty() {
    if env.NArgs() < 3 {
      ker.Panic ("...")
    }
    h, w = env.N(2), env.N(3)
    if h <= 1 { ker.Panic ("...") }
    if w <= 24 { ker.Panic ("...") }
    h_file.Seek (0); h_file.Put (h)
    h_file.Seek (1); h_file.Put (w)
  } else {
    h_file.Seek (0); h = h_file.Get().(uint)
    h_file.Seek (1); w = h_file.Get().(uint)
  }
  h_file.Fin()
  scr.NewWH (2, 24, 8 * w, 16 * h); defer scr.Fin()
  scr.Name ("inferno " + name)
  if ms.Empty() {
    ms.Edit()
  } else {
    ms.Write()
  }
  m := mol.New()
  if pseq.Length (name + mol.Suffix) == 0 {
    m.Construct (name)
  } else {
    m = mol.Constructed (name)
  }
  m.Write (0, 0)
  file := pseq.New (m)
  file.Name (name + ".seq")
  all := set.New (m)
```

```
for i := uint(0); i < file.Num(); i++ {
  file.Seek (i)
  m = file.Get().(mol.Molecule)
  all.Ins (m)
}
if env.E() {
  errh.Hint ("help: F1    end: Esc")
} else {
  errh.Hint ("Hilfe: F1    Ende: Esc")
}
all.Jump (false)
if all.Empty() {
  for {
    m.Clr()
    m.Edit (0, 0)
    if m.Empty() {
      // return ?
    } else {
      all.Ins (m)
      break
    }
  }
}
loop:
for {
  m = all.Get().(mol.Molecule)
  m.Write (0, 0)
  switch c, _ := kbd.Command(); c {
  case kbd.Esc:
    break loop
  case kbd.Help:
    errh.Help (help)
  case kbd.Enter:
    m1 := m.Clone().(Rotator)
    m.Edit (0, 0)
    if ! m.Eq (m1) {
      all.Del()
      all.Put (m)
    }
  case kbd.Up:
    all.Step (false)
  ...
  case kbd.Ins:
    m.Clr()
    m.Edit (0, 0)
    if m.Empty() { ker.Panic ("...") }
    all.Ins (m)
  case kbd.Del:
    if errh.Confirmed() {
      all.Del()
    }
  case kbd.Search:
    ...
  case kbd.Act:
    m.Rotate()
    all.Sort()
  case kbd.Print:
    ms.Print()
    m.Print()
  }
}
errh.DelHint()
file.Clr()
all.Trav (func (a any) { file.Ins (a.(mol.Molecule)) } )
file.Fin()
}
```

Lindenmayer Systems

13

A rose
is a rose
is a rose
is a rose

Gertrude Stein
From Geography and Plays

Abstract

In 1968, the theoretical biologist Aristid Lindenmayer created *L-systems* as an algorithmic formalism for describing developmental processes in biology. In conjunction with computer graphics, they were initially used for modelling simple multicellular organisms, and later for the realistic representation of plants. A key aspect is the recursive *self-similarity* of the structures (take a look at a Romanesco broccoli). In this chapter, we show a series of simple examples.

13.1 System Analysis

The prerequisite for this chapter is the definition of the terms "*alphabet*", "*language*", and "*grammar*"; in particular, those of the *L-systems*, the grammars of the *Lindenmayer systems*.

These basics should enable *graphical interpretations* of L-systems, which, for example, lead to the pictorial representation of *space-filling curves* or—in the sense of Lindenmayer— more interestingly of *plants*.

Particularly impressive, of course, are three-dimensional constructions of plants. However, the algorithm for this should only be limited to very simple examples: for example, plant stems and tree trunks are only represented as lines and leaves only as simple unfilled polygons.

13.1.1 Alphabets, Languages, and Grammars

We understand an *alphabet* to be a finite set A with at least two elements. We refer to the elements of A as *letters* and sequences of letters written one after the other as *words*.

With A^* we denote the set of all words over A, recursively defined by

$$A^0 = \{\varepsilon\} \quad \text{and} \quad A^{n+1} = \{wa \mid w \in A^n \text{ and } a \in A\} \quad \text{for } n \in \mathbb{N}$$

defined, where ε is the *empty word* with $w\varepsilon = w = \varepsilon w$ for all words $w \in A^*$.

Example 13.1 For $A = \{0, 1\}$ a word is a sequence of zeros and ones, thus A^* is the set $\mathbb{N}$ of all natural numbers in binary notation and for $A = \{0, 1, 2, \ldots, 9\}$ in the usual notation.

A *language over the alphabet* A is a set of words with letters from A.

We understand a *grammar* to be a quadruple (A, V, s, P), where

- A is an alphabet;
- V is a non-empty finite set of variables, disjoint from A, $s \in V$ is the *start symbol*; and
- $P \subset (\Sigma^* \setminus A^*) \times (\Sigma^* \setminus \{s\})$ is a finite set of *production rules*.

We refer to the elements of the set $\Sigma = A \cup V$ as *symbols*.

For a production rule $(l, r) \in P$ one usually writes $l \to r$ and says "r is derived from l"; for production rules $(l, r1), (l, r2), \ldots, (l, rn)$ one also writes briefly $l \to r1 \mid r2 \mid \cdots \mid rn$. Here, l contains at least *one* variable, because $l \notin A^*$.

13.1.2 Relationship Between Grammars and Languages

Each grammar (A, V, s, P) generates a language over A. It consists of those words that can be derived from the start symbol s by a sequence of applications of the production rules $(v, w) \in P$. A step in this sequence consists for a production rule $(v, w) \in P$ in that in a sequence of symbols from Σ^*, in which v occurs, this v is replaced by w.

Example 13.2 The grammar (A, V, ε, P) with $A = \{a, b\}$, $V = \{v\}$ and $P = \{(v, avb), (v, \varepsilon)\}$ generates the language $\{a^n b^n \mid n \in \mathbb{N}\}$.

The same language is generated by the grammar with $A = \{a, b\}$, $V = \{A, B, X\}$, $s = X$ and $P = \{X \to ABX, BA \to AB, BX \to b, Bb \to bb, Ab \to ab, AX \to aa\}$.

Conversely, for every language there is a grammar that generates this language. However, we do not go into this very extensive topic about *Turing machines* and *automata* here, as it does not further our purposes for this chapter. Those interested are referred to the specialist literature (e.g., [1]).

13.2 The Grammars of Lindenmayer Systems

These grammars have the form (A, V, ω, P), where

- A is an alphabet,
- V is a set of letters disjoint from A,
- ω is the (non-empty) start word from A^*, called *Axiom*, and
- $P \subset A \times A^*$ is a finite set of production rules.

They are called *L-systems*.

In them, $a \to v$ is written for a production rule $(a, v) \in P$.

In a derivation step for a production rule $(v, w) \in P$ in a word from A^*, in which v occurs, all occurrences of v are simultaneously replaced by w. It is assumed that for each letter $a \in A$ there is exactly one word $v \in A^*$ with $a \to v$.

Example 13.3 The grammar $(a, b, \emptyset, b, (a \to ab), (b \to a))$ generates the language $b, a, ab, aba, abaab, abaababa, abaababaabaab, abaababaabaababaababa, \ldots$.

13.3 Graphical Interpretation of L-Systems

In the following, we adhere closely to [2]. Most of the many examples of graphical interpretations also come from this book.

We first consider L-systems with the alphabet $(F, f, +, -)$. Crucial for the following is that it is possible to interpret each such L-system as a graphic.

A *state* of this graphic is a triple (x, y, α), where

- $(x, y) \in \mathbb{R}^2$ is a position in the plane and
- α is an angle.

This includes a starting angle $\alpha \in \mathbb{R}$, a rotation angle $\delta \in \mathbb{R}$ (both given in degrees), and the number $n \in \mathbb{N}$ of application steps of production rules.

We change the state of the L-system for the given symbols as follows:

- F: We move forward by one step (of length 1), which transforms the position (x, y) into the position $(x + \cos\alpha, y + \sin\alpha)$, and draw a line between the old and new positions.
- f: We do the same as with F, but without drawing a line.
- $+$: We turn to the left by the angle δ, which transforms the state (x, y, α) into the state $(x, y, \alpha + \delta)$.
- $-$: We turn to the right by the angle δ, which transforms the state (x, y, α) into the state $(x, y, \alpha - \delta)$.

13.3.1 The Koch Islands

As a first example of a graphical interpretation of an L-system, we consider the system with

- the axiom $\omega =$ F+F+F+F$\in A^*$,
- the only production rule F$\to$F-F+F+FF-F-F+F.
- the starting angle $\alpha = 0°$, and
- the rotation angle $\delta = 90°$.

In one application step, the symbol sequence

F−F+F+FF−F−F+F−F+F+FF−F−F+F−F+F+FF−F−F+F−F+F+FF−F−F+F.

is obtained. We refrain from specifying symbol sequences that result from more steps—that would only be boring.

Much prettier are the graphical interpretations: In Figures 13.1, 13.2, 13.3, 13.4 to 13.5, the Koch Island for 0 to 4 application steps can be seen.

13.3.2 The Islands and Lakes

As a second example, we consider the system with

Fig. 13.1 Koch Island: start

13.3 Graphical Interpretation of L-Systems

Fig. 13.2 Koch Island after 1 application step

Fig. 13.3 Koch Island after 2 application steps

Fig. 13.4 Koch Island after 3 application steps

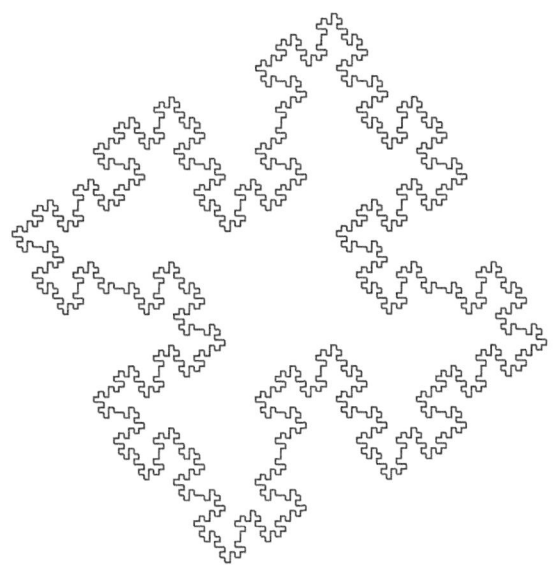

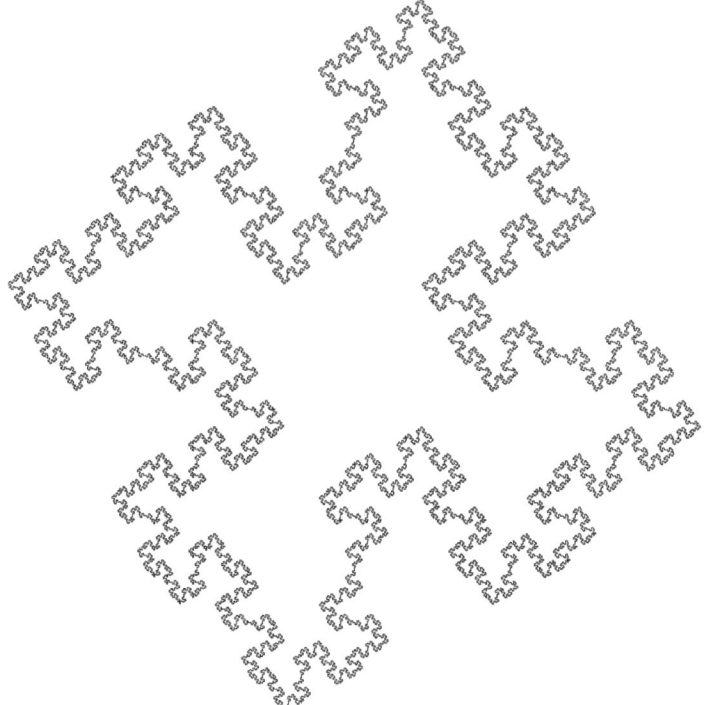

Fig. 13.5 Koch Island after 4 application steps

- the axiom $\omega = \text{rrF+oF+gF+bF}$,
- the production rules $\text{F} \rightarrow \text{F+f-FF+F+FF+Ff+FF-f+FF-F-FF-Ff-FFF}$,
- $\text{f} \rightarrow \text{ffffff}$, and
- the rotation angle $90°$.

Figure 13.6 shows the system after two application steps.

13.3.3 The Pavement

This example is almost like something in the next section.

- the axiom $\omega = \text{F-F-F-F}$,
- the production rule $\text{F} \rightarrow \text{F -> rFF-oF+gF-cF-bFF}$, and
- the rotation angle $90°$.

Figure 13.7 shows it after two application steps and Fig. 13.8 after five application steps.

13.3 Graphical Interpretation of L-Systems

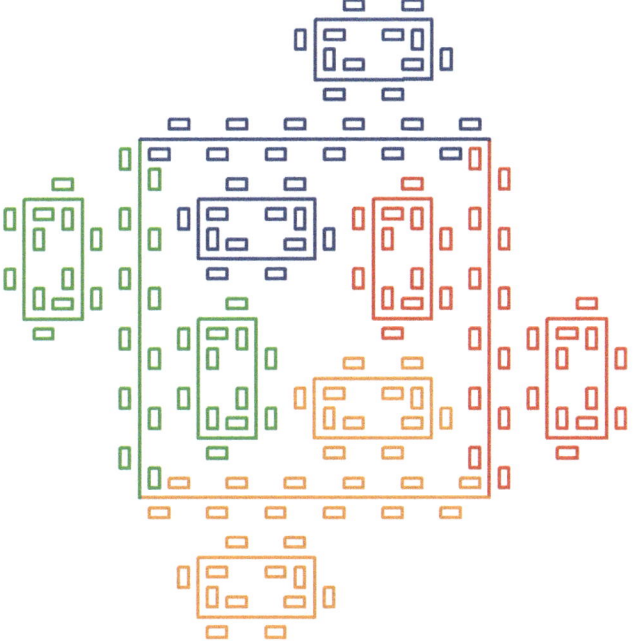

Fig. 13.6 Islands and lakes

Fig. 13.7 Pavement after two application steps

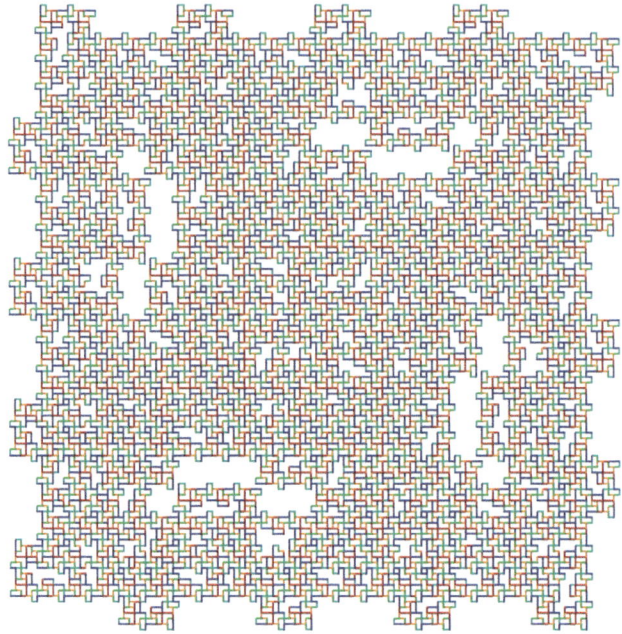

Fig. 13.8 Pavement after five application steps

13.3.4 Space-Filling Curves

Our next examples provide *space-filling curves*.

We are expanding our alphabet to include lowercase letters as symbols for colours:

- n for brown,
- r for red and l for light red,
- o for orange,
- g for green and d for dark green,
- c for cyan,
- e for light blue and b for blue,
- m for magenta, and
- y for grey.

13.3.4.1 The Hilbert Curves

First, we consider the system with

- the variables X and Y,
- the axiom $\omega = $ X,

Fig. 13.9 Hilbert curve after two application steps

Fig. 13.10 Hilbert curve after four application steps

- the production rules X→X→+rYrF-XgFX-rFY+, and
- the production rules X→X→+YrF-XgFX-rFY+ and X→Y→-XrF+YgFY+rFX-.

Figures 13.9, 13.10, and 13.11 show the Hilbert curve for two, four, and seven application steps.

13.3.4.2 The Peano Curves
The next example with

- the variables L and R,
- the axiom ω = -L∈ A^*, and
- the production rules L→F→LF+RFR+FL-F-LFLFL-FRFR+ and R→-LFLF+RFRFR+F+RF-LFL-FR

provides for $n = 2$ and $n = 4$ Figures 13.12 and 13.13.

13.3.4.3 The Barrel Curves
The last example of real space-filling curves with

- the variables L and R,
- the axiom ω = -L∈ A^*, and
- the production rules L→F→LF+RFR+FL-F-LFLFL-FRFR+ and R→-LFLF+RFRFR+F+RF-LFL-FR

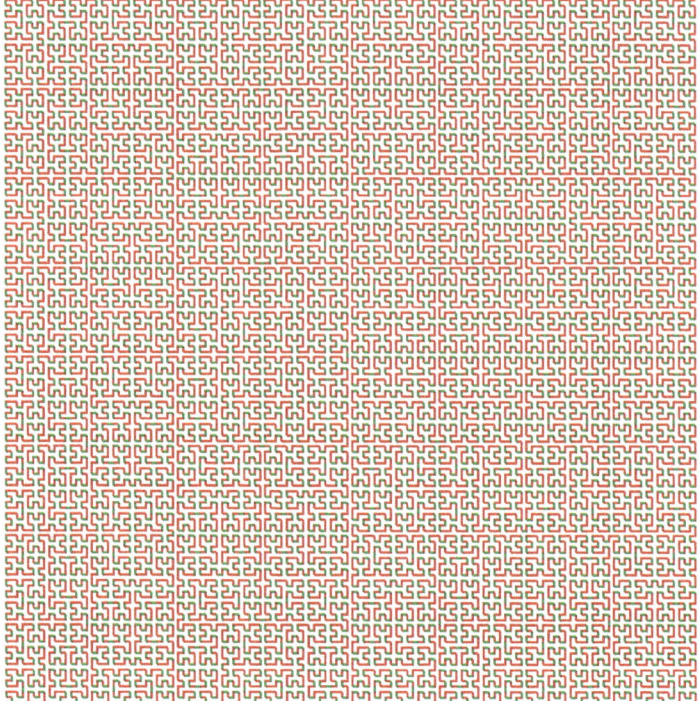

Fig. 13.11 Hilbert curve after seven application steps

Fig. 13.12 Peano-curv after
two application steps

provides for $n = 2$ and $n = 4$ the barrel curves in Figures 13.14 and 13.15.

13.3.4.4 The Sierpinski Curves
The file

- the axiom ω = F+F+F+F and
- the production rule F→F→X -> XF-F+F-XF+F+XF-F+F-X.

provides for $n = 2$ and $n = 6$ Figure 13.16, resp. 13.17.

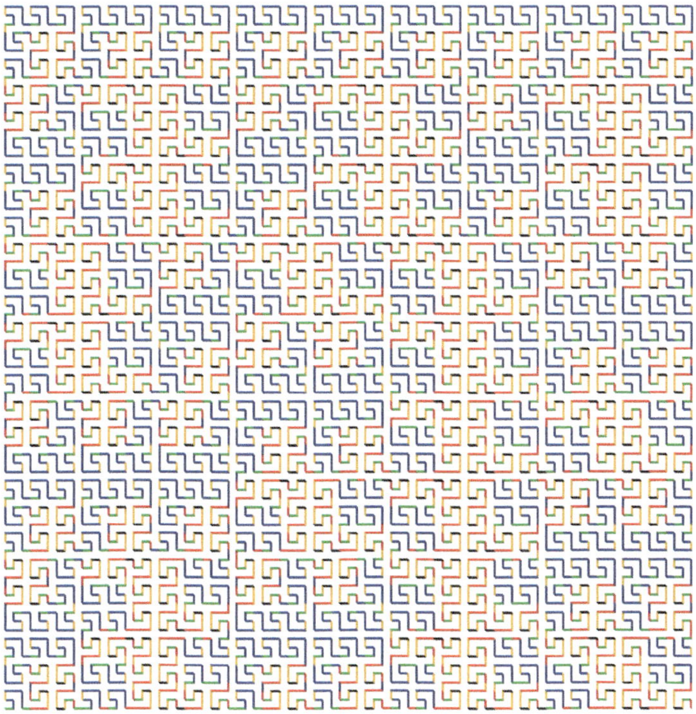

13.3 Graphical Interpretation of L-Systems

Fig. 13.13 Peano-curv after four application steps

Fig. 13.14 Barrel curve after two application steps

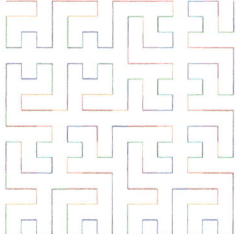

13.3.5 Extensions of the Alphabet of L-Systems

For *forward steps* (with or without drawing a line between the start and end position of the step), we have introduced the letters "F" and "f" in Sect. 13.3, and for *left* and *right turns* by the given angle of rotation, the rotation letters "+" and "-".

In addition, there is the reversal letter "|", with which the direction of the step (by 180°) is reversed.

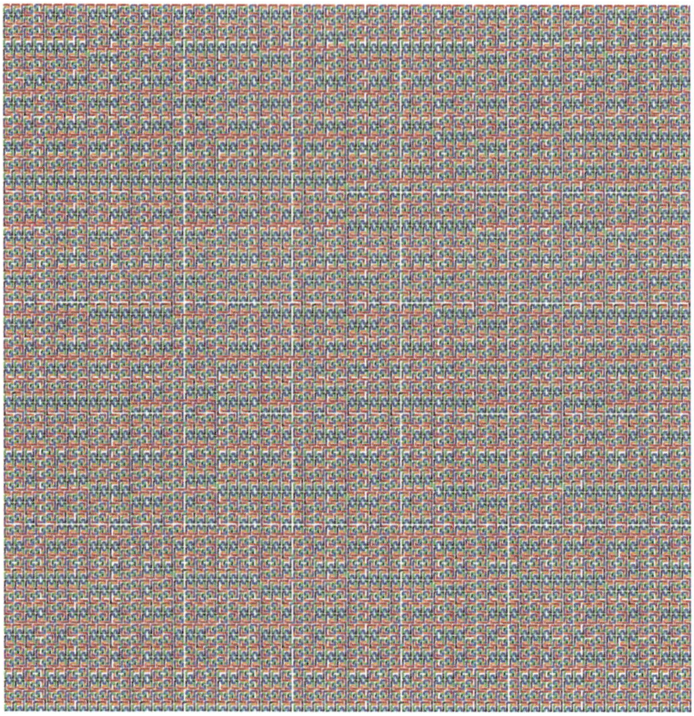

Fig. 13.15 Barrel curve after four application steps

Fig. 13.16 Sierpinski curve
after two application steps

The ls files can be annotated line by line for documentation purposes. To do this, the corresponding lines must begin with the symbol "%"; the content of these lines does not contribute to the construction of an L-system.

13.3.5.1 Branches

The two following symbols are particularly important as they allow branches, because they enable the creation of plant illustrations—the intended goal of Lindenmayer:

- [for the start of a branch and
-] for the end of a branch.

13.3 Graphical Interpretation of L-Systems   249

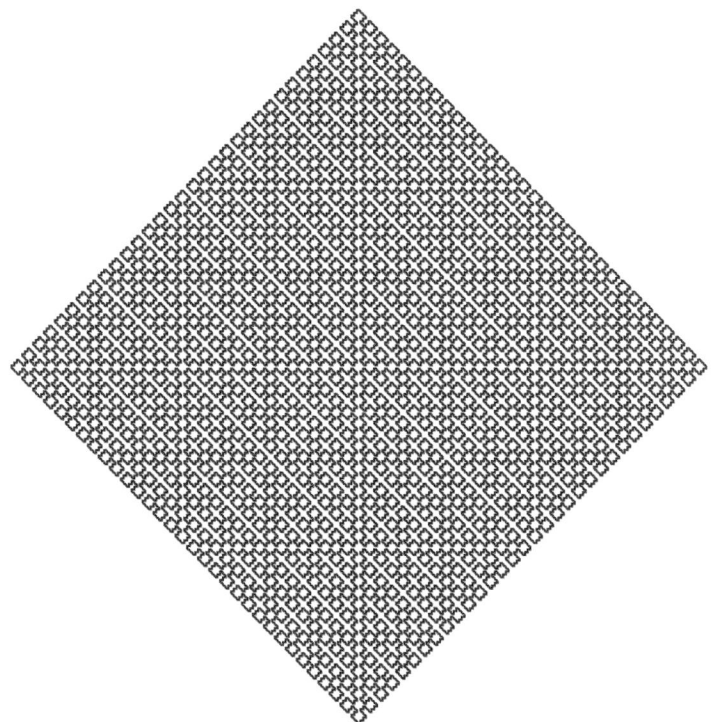

Fig. 13.17 Sierpinski curve after six application steps

We will show six nice examples of this.

The first two of these with the definitions

```
270
gF
F  -> dF[+nF]gF[-nF]dF
25
5
```

and

```
270
F
F  -> gF[+oF]dF[-rF][gF]
25
5
```

in Fig. 13.18.

Fig. 13.18 Two herbs

The third example

```
270
gF
F  ->  nFF-g[-F+F+F]+g[+F-F-F]
22.5
4
```

provides a bush (s. Fig. 13.19).

The two next examples with the definitions

```
270
X
X  ->  gF[+gX]dF[-X]+X
F  ->  fdFgF
20
8
```

and

```
270
X
X  ->  lF[+X][-X]mFX
F  ->  nFgF
25
7
```

13.3 Graphical Interpretation of L-Systems

Fig. 13.19 A bush

are shown in Fig. 13.20.
  With
```
270
X
X  -> gF-[[X]+X]+oF[+gFX]-X
F  -> FdF
35
6
```
another herb is defined (s. Fig. 13.21).

13.3.6 Three-Dimensional L-Systems

We can also construct three-dimensional L-systems. To do this, the alphabet of the L-systems is extended by the following symbols:

- {   start of a polygon,
- }   end of a polygon,
- _   incline forward,
- ^   incline backward,

Fig. 13.20 Two herbs

- / tilt to the left, and
- \ tilt to the right.

Of course, no *three-dimensional* models can be depicted on the "two-dimensional" paper of this book. For this reason, we show a second view for each example.

A real impression of the models is provided by the program lsys. It relies on the function Go from the package μU for the representation of 3D scenes. With it, you can

- move left, right, up, down, closer to the centre of the model or further away from it;
- turn and tilt left and right;
- incline forward and backward; and
- rotate and tilt the model around the focus.

13.3.6.1 Three-Dimensional Hilbert Curve

The examples of the space-filling Hilbert curves from Sect. 13.3.4.1 can be generalized to three dimensions. The file

13.3 Graphical Interpretation of L-Systems

Fig. 13.21 Another herb

```
X
X  ->  ^\XrF^\XoFX-gF^//XcFX_bF+//mXFkX-yF/X-/
3
```

delivers the three-dimensional Hilbert curve in Fig. 13.22 after three application steps.

Figure 13.23 shows another view of the same example.

13.3.6.2 Three-Dimensional Plants

The file

```
0
F
F  ->  nFF-g[-F+F+F]+/g[+F-F-F]
22.5
4
```

provides a three-dimensional *bush*. We show it in Figures 13.24 and 13.25.

Also, *tree-like structures* can be created.

The file

```
90
A
A  ->  dF[_gFL A]/////R[_gFL A]/////////^[_gFL A]
F  ->  S //// nF
S  ->  F L
L  ->  [^^{-f+f+f-|-f+f+f]
```

Fig. 13.22 Three-dimensional Hilbert curve

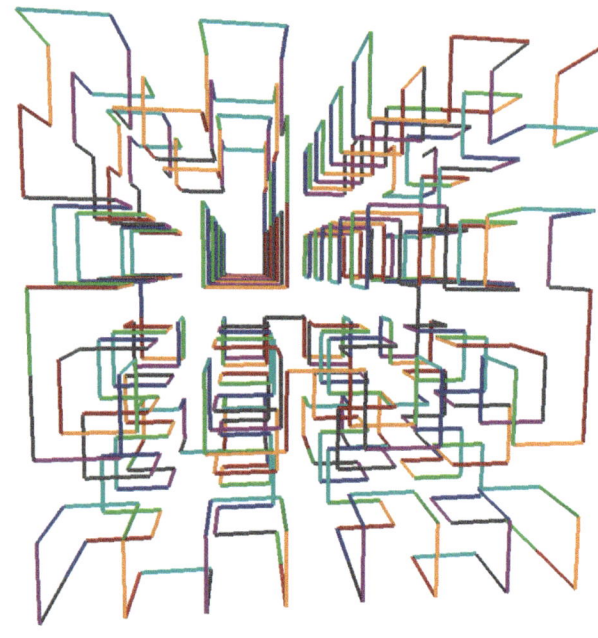

Fig. 13.23 Another view of the three-dimensional Hilbert curve

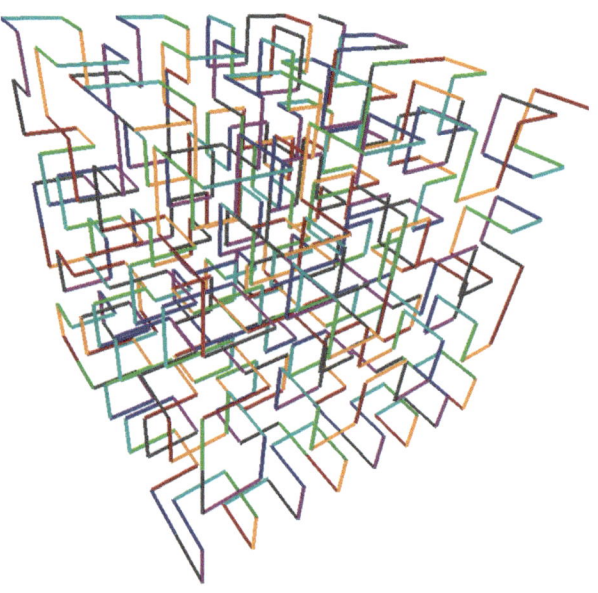

13.3 Graphical Interpretation of L-Systems

Fig. 13.24 A three-dimensional bush

Fig. 13.25 Another view of
the three-dimensional bush

Fig. 13.26 Simple three-dimensional tree

```
22.5
5
```

provides a simple model of a three-dimensional *tree* (see Fig. 13.26).

With the definition

```
nnnT
T   -> R[___+T]-[--^^gLT][^^///gT]R[^++///oL]^^R[-___T]/R[+^^//gL]
R   -> F[--///^gL][^^^n//^^L][+^///gL][+//^^rL]F
L   -> [g+FX-FX-FX+|+FX-FX-FX][^^///[g+FX-FX-FX|+FX-FX-FX]]
FX  -> FX
F   -> FF
22.5
3
```

results in a somewhat less abstract tree (see Figures 13.27 and 13.28).

Figure 13.29 shows two views of a three-dimensional model of a simple flower, based on the file

```
F
F   -> gP
P   -> I+[P+L]--//[--B]I[++B]-[PL]++PL    % plant
I   -> dFS[//___B][//^^B]FS               % branch
S   -> SFS                                % "growing" Stil
B   -> g[+FX-FXFX-FX+|+FX-FXFX-FX]        % leaf
L   -> [___rE/W///W////W////W////W]       % blossom
E   -> FF                                 % stalk
W   -> [o^F][e____-FX+FX|-FX+FX]          % blossom
FX  -> FX
18
4
```

13.3 Graphical Interpretation of L-Systems

Fig. 13.27 Three-dimensional tree

Fig. 13.28 Other view of the three-dimensional tree

Fig. 13.29 Simple three-dimensional flower

Here is the penultimate example: The file

```
F
F -> nFF-u[-F+F+F]+/u[+F-F-F]
22.5
4
```

provides a three-dimensional grass-like plant (see Figures 13.30 and 13.31).

As a final example, we show with the file

```
270
F
F -> gF[/+rF][//+lF][///+rF]yF[\-oF][//+oF][\\\\-lF]
30
3
```

in Fig. 13.32 another fantasy plant.
in Fig. 13.32 another fantasy plant.
F -> gF[/+rF][//+lF][///+rF]yF[oF][//+oF][
-lF] 30 3

Fig. 13.32 three-dimensional fantasy plant

13.4 System Architecture

Fig. 13.30 Three-dimensional grass plant

Fig. 13.31 Three-dimensional grass plant from above

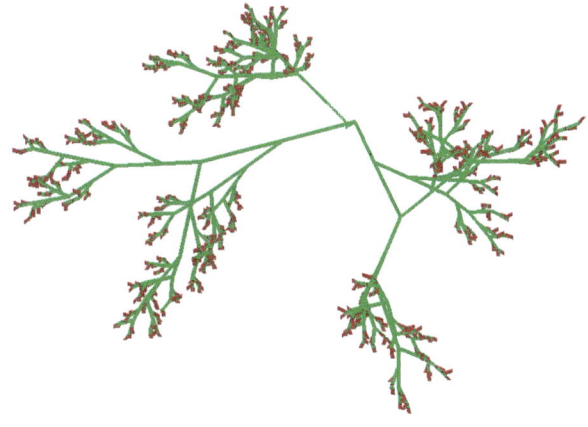

13.4 System Architecture

The essential parts of the system are

- the grammar and
- the ongoing state during the construction of an L-system.

Fig. 13.32 Three-dimensional fantasy plant

13.4.1 The Objects of the System

This results in the following objects:

- the *grammar* as an abstract data object,
- an abstract data type for managing the ongoing state during the construction of an L-system—i.e., the *coordinates* and *directions*.

They can be found in the packages

- lsys/grammar,
- μU, which manages an "eye point", a "focus", and a normalized orthogonal triad "(right, front, top)" in the $\mathbb{R}^3$.

For managing the ongoing state, a stack for pairs of symbols and the respective position of the symbol in a production rule is also needed as an abstract data object. It forms the package

- lsys/symstk with the subpackage and
- lsys/symstk/pair.

13.4.2 Component Hierarchy

The dependencies of the packages are shown in Fig. 13.33, where the top layer consists only of the main program `lsys.go`, which uses the three underlying packages.

13.5 User Manual

First, a file with the suffix ".ls" must be created using any text editor, which defines an L-system. The lines of this file must meet a series of requirements, which we will explain in detail in the following section.

13.5.1 Creation of an L-System

The lines of the L-system files must meet the following conditions:

- The first line can contain a natural number < 360 that specifies the starting angle of the geometric interpretation. If there is no first line with a number, the starting angle is 90°.
- The axiom must be in the next (or possibly the first) line. This line may only contain symbols.
- The subsequent lines contain the production rules (at least *one* must be specified). Only symbols may appear in this line as well.
- The penultimate line must consist of a natural number < 360. It specifies the angle by which the direction of the step changes when a *rotation symbol* appears on the right side of a production rule.
- The last line must contain the number of application steps, a natural number < 26.

13.5.2 System Operation

It is incredibly simple:

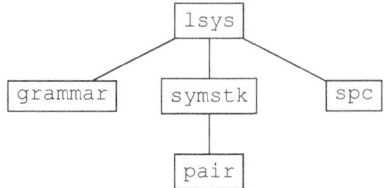

Fig. 13.33 System architecture of the L-System

If the edited file meets these requirements, the geometric interpretation appears on the screen after calling the program `lsysi` with the name of the file as an argument (even without the suffix `.ls`)—however, for three-dimensional systems, this only happens on a graphical interface.

If this is not the case, an appropriate error message will appear.

If the name of an `ls`-file that does not exist is passed as a parameter to the program call, the call has no effect.

13.6 Construction

13.6.1 Specification of the Library Packages

Below, we show the specifications of the three packages used.

13.6.1.1 Grammar

The specification of the abstract data type in the package `lsys/grammar` is as follows:

```
package grammar
import "µU/col"

const (
  MaxL =    4
  MaxR =   80
  Comment       = '%'
  Step          = 'F'
  YetiStep      = 'f'
  TurnLeft      = '+' // around z-axis
  TurnRight     = '-'
  Invert        = '|'
  TiltDown      = '_' // around x-axis
  TiltUp        = '^'
  RollLeft      = byte(92) // '\' // around y-axis
  RollRight     = '/'
  BranchStart   = '['
  BranchEnd     = ']'
  PolygonStart  = '{'
  PolygonEnd    = '}'
)
type
  Symbol = byte
var (
  StartColour col.Colour
  Startword string
  Startangle, Turnangle float64
  NumIterations uint
  colours = []col.Colour {col.Brown(),     // n
                          col.Red(),       // r
                          col.LightRed(),  // l
                          ...
                         }
)

func Initialize (s string) { initialize(s) }

func IsColour (s Symbol) (col.Colour, bool) { return isColour(s) }
```

13.6 Construction

```
// Pre: s is not empty.
// Returns true, iff there is a rule with a left side starting with s.
func ExRule (s string) bool { return exRule(s) }

// Pre: There is at most one rule with s as left side.
// Returns the right side of the rule with left side s,
// if such a rule exists; otherwise "".
func Derivation (s string) string { return derivation(s) }
```

13.6.1.2 Symbol Stack

The specification of the symbol stack is directly derived from the specification of the general stack μU in the microuniverse:

```
package symstk // A stack of pairs (byte, uint); initially empty.

type Symbol = byte

// (s, i) is pushed onto the stack.
func Push (s Symbol, i uint) { push(s,i) }

// Returns true, iff x is empty.
func Empty() bool { return empty() }

// Pre: x is not empty.
// Returns the pair on top of x. That pair is now removed from x.
func Pop() (Symbol, uint) { return pop() }
```

13.6.1.3 Management of the Current Position and Directions

This is done during the construction of the geometric interpretation of an L-system in the package μU. Here is its specification:

```
package spc
// The package maintains the following 5 vectors:
// origin, focus and an orthogonal right-handed trihedron
// consisting of the 3 vectors (right, front, top) with
// len(right) = len(front) = len(top) = 1,
// s.t. front = focus - origin normed to len 1.
// Maintains furthermore a stack of the trihedron-vectors.
// origin = (ox, oy, oz), focus = (fx, fy, fz), top = (tx, ty, tz),
// front = focus - origin normed to len 1 and
// right = vector-product of front and top.
func Set (ox, oy, oz, fx, fy, fz, tx, ty, tz float64) {
set (ox,oy,oz,fx,fy,fz,tx,ty,tz) }

// Returns the coordinates of origin, focus and top.
func GetOrigin() (float64, float64, float64) { return getOrigin() }
func GetFocus()  (float64, float64, float64) { return getFocus() }
func GetRight()  (float64, float64, float64) { return getRight() }
func GetFront()  (float64, float64, float64) { return getFront() }
func GetTop()    (float64, float64, float64) { return getTop() }

// origin is moved in direction Right/Front/Top by distance d,
func MoveRight (d float64) { moveR(d) }
func MoveFront (d float64) { moveF(d) }
func MoveTop   (d float64) { moveT(d) }

// origin and focus are moved in direction Right/Front/Top by
```

```
distance d,
func Move1Right (d float64) { move1R(d) }
func Move1Front (d float64) { move1F(d) }
func Move1Top (d float64) { move1T(d) }

// front is rotated around right by angle a, top is adjusted.
func Tilt (a float64) { tilt(a) }

// top is rotated around front by angle a, right is adjusted.
func Roll (a float64) { roll(a) }

// right is rotated around top by angle a, front is adjusted.
func Turn (a float64) { turn(a) }

// The trihedron is rotated around the vector right by angle a.
func TurnAroundFocusR (a float64) { turnAroundFocusR(a) }

// The trihedron is rotated around the vector top by angle a.
func TurnAroundFocusT (a float64) { turnAroundFocusT(a) }

// Returns true, iff the stack is empty.
func Empty() bool { return empty() }

// origin, focus and top are pushed onto the stack.
func Push() { push() }

// origin, focus and top are popped from the stack
// and front and right are computed to maintain the invariants.
func Pop() { pop() }
```

13.6.2 Implementation of the Packages

The main program distinguishes between the construction of two- and three-dimensional L-systems, which is determined by whether Tilt or Roll symbols appear in the ls-file.

13.6.2.1 Main Program

We only show the implementation of one of the most important functions in the main program, the function step, which changes the state of the L-system depending on the actual symbol:

```
func step (s Symbol) {
  switch s {
  case g.Step:
    x0, y0, z0 = spc.GetOrigin()
    spc.Move1Front (1)
    x, y, z = spc.GetOrigin()
    ...
  case g.YetiStep:
    x0, y0, z0 = spc.GetOrigin()
    spc.Move1Front (1)
    x, y, z = spc.GetOrigin()
  case g.TurnLeft:
    spc.Turn (delta)
    ox, oy, _ := spc.GetOrigin()
    alpha = arctan (ox, oy)
  case g.TurnRight:
    spc.Turn (-delta)
```

```
    ox, oy, _ := spc.GetOrigin()
    alpha = arctan (ox, oy)
  case g.Invert:
    spc.Turn (180)
  case g.TiltDown:
    spc.Tilt (delta)
  case g.TiltUp:
    spc.Tilt (-delta)
  case g.RollLeft:
    spc.Roll (delta)
  case g.RollRight:
    spc.Roll (-delta)
  case g.BranchStart:
    spc.Push()
  case g.BranchEnd:
    spc.Pop()
  default:
    ...
  }
}
```

13.6.2.2 Grammar

The implementation of this package mainly consists of examining the passed ls-file with the verification of the symbols and the construction of the production rules, so that they can be processed by the main program.

13.6.2.3 Symbol Stack

Its implementation essentially only consists of accessing the stack package μU in the microuniverse. It is only used during the execution of the function execute for measuring and creating the two-dimensional graphics and for creating the three-dimensional graphics as openGL-Constructs needed.

13.6.2.4 Space

The implementation makes intensive use of the vector package μUf the microuniverse.

References

1. Chomsky, N.: Three models for the description of language. In: IRE Transactions in Information Theory **2**, 113–126 (1956). https://www.doi.org/
2. Prusinkiewicz, P., Lindenmayer, A.: The Algorithmic Beauty of Plants. Springer, New York (1990). https://doi.org/10.1007/978-1-4613-8476-2

Rail 14

Abstract

The operation of a train station essentially consists of the construction of *routes* for train journeys. *Routes* are sequences of *blocks*; blocks are *track sections*, on which only one train may be present at a time; and track sections are, for example, *tracks*, *switches*, and *double crossover switches* and *buffer stops*. This hierarchy thus provides a nice example of object-based programming. The project consists of simulating a pushbutton control panel for a train station and has a distributed aspect: the simulation of train traffic between several stations.

> Sprachlich bedeutet Eisenbahn ganz allgemein eine Bahn von Eisen zwecks Bewegung von Gegenständen auf derselben. Verknüpft man diesen Wortlaut mit dem Gesetzeszweck, und erwägt man, daß die eigenartige Nützlichkeit und gleichzeitig Gefährlichkeit des metallischen Transportgrundes, in der (durch dessen Konsistenz, sowie durch dessen, das Hinderniß der Reibung vermindernde Formation und Glätte gegebenen) Möglichkeit besteht, große Gewichtsmassen auf jenem Grunde fortzubewegen und eine verhältnißmäßig bedeutende Geschwindigkeit der Transportbewegung zu erzeugen, so gelangt man im Geiste des Gesetzes zu keiner engeren Bestimmung jener sprachlichen Bedeutung des Wortes Eisenbahn, um den Begriff eines Eisenbahnunternehmens im Sinne des Gesetzes zu gewinnen, als derjenigen: Ein Unternehmen, gerichtet auf wiederholte Fortbewegung von Personen oder Sachen über nicht ganz unbedeutende Raumstrecken auf metallener Grundlage, welche durch ihre Konsistenz, Konstruktion und Glätte den Transport großer Gewichtsmassen, beziehungsweise die Erzielung einer verhältnißmäßig bedeutenden Schnelligkeit der Transportbewegung zu ermöglichen bestimmt ist, und durch diese Eigenart in Verbindung mit den außerdem zur Erzeugung der Transportbewegung benutzten Naturkräften (Dampf, Elektricität, thierischer oder menschlicher Muskelthätigkeit, bei geneigter Ebene der Bahn auch schon der eigenen Schwere der Transportgefäße und deren Ladung, u.s.w.) bei dem Betriebe des Unternehmens auf derselben eine verhältnißmäßig gewaltige (je nach den Umständen nur in bezweckter Weise nützliche, oder auch Menschenleben vernichtende und die menschliche Gesundheit verletzende Wirkung zu erzeugen fähig ist.
>
> Civil Senate of the Reich Court Leipzig from March 17, 1879
> Meaning of the expressions
> Operation of a railway and operating company
> in the § 1 of the Reich Liability Act
> Decisions of the Reich Court in Civil Matters I (1880), 247–252

Essential parts of the concept and approaches of the Bahn project are based on the work results in a basic computer science course at the Rückert-Gymnasium in Berlin-Schöneberg many decades ago. It was about the simulation of an "electronic interlocking" in a station; the occasion was the introduction of the first microcomputer-controlled interlocking by Siemens in the Berlin U-Bahn Station Uhlandstraße.

Since Pascal was the only programming language available to us at the time, I was able—knowing the works of Parnas—to achieve only initially a separation between specification and implementation using the forward declarations. Due to the lack of graphics on our ASCII terminals, we represented the cells (see Sect. 14.3.1.1) with

- the minus sign "-",
- the underscore "_",
- the slash "/",
- the backslash "\", and
- the separator " | ".

I set this task to the participants of some teacher training courses in computer science at the Free University of Berlin a few years later; object-based programming was possible with Modula-2 at that time.

Various extensions planned at the time as options were developed by me using the microuniverse and the various parts of the documentation were completely revised.

In particular, the system has been expanded to a small network of six stations.

Thanks are due to

- my former students, who first built a Brio railway in the computer room to get clarity about what the essence of, for example, switches or routes, then devised a clever method for implementing lock variables, as our trains were started concurrently from several terminals, later rediscovered depth-first search in graphs, and finally found shortest routes for trains by sorting through merging in sets of lists of blocks (thus mastering very demanding content from the field of algorithms and data structures);
- the course participants of the teacher training, who had continuously driven the development; and
- Mr. Dipl.-Ing. Norbert Ritter (then project manager at the Berlin Transport Company), who showed us the new digital interlocking in the U-Bahn Station, and Mr. Prof. Rolf Schädlich from the TFH, who showed us a dismantled mechanical interlocking of the railway at the TFH.

14.1 System Analysis

It is about a program system that supports the *train service management* (EBO § 47 (1) 4) in a given *station* (EBO § 4 (2)) in their task to ensure the safety of the *train operation*, (EBO § 47 (2) and (3)): The system should provide all information necessary for the issuance of *driving orders* required.

This includes the display of the track diagrams of the stations, i.e.:

- of all tracks with their numbers;
- of all switches, double crossover switches, and signals with their positions; and
- the occupancy reports by stationary and moving trains.

A driving order is triggered by entering the start and destination tracks. The system then constructs all possible routes, selects the one with the fewest switches as the route, secures it by switching the used switches and signals, and marks it as occupied. The driving order is issued by setting the driving signal.

For the assembly or division of trains, shunting movements into an occupied destination track are possible; moreover, it cannot be assumed that the starting track will become free after the departure of a shunting section. A model with such shunting possibilities would have to keep track of the composition of all trains from their components (locomotive(s) and wagons including their relative positions to each other), otherwise meaningless driving orders are possible. Due to the *significant* complexity of a solution that includes all these factors, *shunting* movements are not modelled—we limit ourselves to train journeys.

Only a *phantom* is considered: The train journeys are only *simulated on the screen*.

Trains leaving the station go to the neighbouring station. New trains can appear on the station's entrance tracks and must be "picked up" there.

This allows the task of the system to be formulated as follows:

It is about the *simulation of pushbutton interlocking based on the track diagram display* and the *synchronization of train traffic between stations*.

14.1.1 Basic Concepts of Railway Technology

A *block* (EBO § 4 (3)) is a section of track at whose ends the continuation of a train in the permitted *directions of travel* is controlled (released or blocked) by signals. The directions of travel for which a block is permitted depend on its function, e.g., on double-track main lines, the blocks are usually only permitted for *one direction of travel* (in Germany *right-hand traffic*) (EBO § 38).

The *block protection* is therefore a central safety concept in train operation: Within a block, there can always be only *one* train.

Exiting the block and thus entering the next block is only possible if there is no train in the next block.

In a *station*, each longer section of track between two switches is to be treated like a block. There can always be at most one train on it; exit signals are at its ends. Switches are also to be considered as blocks in this sense, as they can only be used by one train for obvious reasons.

Roads are in this sense *sequences of blocks*; *tracks* are routes that are signal-technically secured.

In order to identify the tracks for the registration of journeys and switching of *roads*, the usual scheme for track numbering is used: parallel main tracks are counted from the reception building, merging tracks are counted in both directions in full decades, with the last digit of their number matching the number of the main track they continue.

The main tracks for through traffic may only be used by trains in the direction of regular operation.

The terms *"with the kilometerage"* (i.e., in ascending order of the kilometre stones on the railway line on which the station is located) and *"against the kilometerage"* are agreed upon as directions of travel; in this sense, all tracks of a station are oriented—reversing loops are not permitted.

Switches in this sense are either *with* the or *against* the kilometerage branched. They can (seen in the direction of branching) have the positions *right* or *left*, abstracting from the (railway—technically important) information that switches have a straight and a branching branch. *Double crossover switches* are branched in both directions.

Switch numbers are assigned *in the direction of the kilometerage*, with switches that directly connect parallel tracks having consecutive numbers. At individual station sections (e.g., at the station heads), new decades start each time.

In addition to the branching of the tracks, the switches have an important function for the safety of train operation: occupied or travelled track sections must be protected against a lateral entry by rejecting switch positions in the direction of a parallel track (*flank protection*).

The representation of the *signals* is part of the simulation of station operation. Both *main* and *protection signals* (ESO B I. and VII.) are taken into account. The inclusion of *advance signals* (ESO B II. and III.) as well as other signals (additional signals, slow driving signals, shunting signals, etc.) is reserved for a later expansion stage.

14.1.2 Sources

The essential regulations of the Deutsche Bahn for train traffic are the

- Railway Construction and Operating Regulations (EBO) [1],
- Driving Service Regulations [2], and
- Railway Signal Regulations (ESO) 1959 [3]

of the Deutsche Bahn.

14.1.2.1 Excerpts from the Railway Construction and Operating Regulations

§ 4 Definitions

(1) Railway facilities are all properties, structures, and other facilities of a railway, which are necessary for the handling or securing of passenger or freight traffic on the rail, considering the local conditions. This also includes auxiliary operating facilities and other facilities of a railway that enable or promote loading and unloading as well as access and exit. There are railway facilities of the stations, the open line and other railway facilities. Vehicles are not part of the railway facilities.
(2) Stations are railway facilities with at least one switch, where trains may start, end, evade, or turn. The entry signals or trapezoid boards generally serve as the boundary between the stations and the open line, otherwise the entry switches.
(3) Block sections are track sections into which a train may only enter if they are free of vehicles.
(11) Main tracks are the tracks regularly used by trains. Continuous main tracks are the main tracks of the open line and their continuation in the stations. All other tracks are sidings.

§ 34 Definition, type and length of trains

(1) Trains are the units moving onto the open line consisting of standard vehicles, moved by mechanical power, and individually moving locomotives.

§ 38 Driving order

On double-track railways, the right side is to be used for driving. This can be deviated from
1. in stations and when introducing track lines into stations,
2. …

§ 39 Train sequence

(4) The entry, exit, or passage of a train may only be permitted if its route is clear. …

14.1.2.2 Excerpts from the Railway Signal Regulations

I. Main signals (Hp)

(6) Main signals are used as entry signals, exit signals, intermediate signals, block signals,
…

(10) Main signals indicate whether the subsequent track section may be used. The main signals Hp 0, Hp 1, and Hp 2 only apply to train journeys, but not to shunting movements.

Hp 0: Train stop

Light signal: One red light.

Hp 1: Go

Light signal: One green light.

Hp 2: Slow movement

Light signal: One green and vertically below it a yellow light.

The screen displays the station in the form of a track diagram display, i.e., the schematic representation of all tracks, switches, and double crossover switches and the position of all signals. The dynamic part of the display consists at any given time of the representation of the current state of its variable components: the positions of all switches and signals and the occupancy reports of all tracks—differentiated according to whether they are occupied by stationary trains or by issued driving orders or moving trains. When a train has left a track, the occupancy report is withdrawn. (In reality, track clearance detectors are used, devices on the tracks that register the state, for example, by counting the passing axles.)

The train dispatcher requests a route for a train journey by pressing buttons on the track display that indicate the start and the destination track for the journey, thereby giving the track display the function of a pushbutton interlocking. Start and destination can in principle only be tracks that are marked by numbers. Switches can only be used for transit; trains are not allowed to stop on them. A journey in this sense is a movement from the start to the destination track in exactly one direction. If a driving order cannot be executed because there is no or no free route from the start to the destination track, or because the destination track is occupied, corresponding messages are issued. The system then checks whether a route is available for the journey. If a driving order can be executed on more than one route, the system selects the one with the fewest switches to be traversed. The system then secures the route by switching the switches and signals for the route, marks all tracks and switches on the route as occupied, and updates the track display in the screen window. The driving order is issued to the train control by setting the corresponding signal to drive or slow drive position. When a train has arrived on an entry track from the neighbouring station, it must be "picked up" by the train dispatcher, i.e., a driving order (from the entry track as the start track) to a destination track in the station must be issued. If an exit track is the destination of a journey, the train disappears in the direction of the following station.

14.1.3 Track Diagram Display

The screen displays the station in the form of a *track diagram display*, i.e., the schematic representation of all tracks, switches and double crossover switches, and the position of all signals.

The dynamic part of the display consists at any given time of the representation of the current state of its variable components: the *positions of all switches and signals* and the *occupancy reports of all tracks*—differentiated according to whether they are occupied by stationary trains or by issued driving orders or moving trains. When a train has left a track, the occupancy report is withdrawn. (In reality, *track clearance detectors* are used, devices on the tracks that register the state, for example, by counting the passing axles.)

14.1.4 Driving Orders

The train dispatcher requests a route for a train journey by pressing buttons on the track display that indicate the *start* and the *destination track* for the journey, thereby giving the track display the function of a *pushbutton interlocking*.

Start and destination can, in principle, only be tracks that are marked by numbers. Switches can only be used for transit; trains are not allowed to stop on them.

A journey in this sense is a movement from the start to the destination track in *exactly one direction*.

If a driving order cannot be executed because there is *no* or no *free* route from the start to the destination track, or because the *destination track is occupied*, corresponding messages are issued.

The system then checks whether a road is available for the journey.

If a driving order can be executed on more than one road, the system selects the one with the fewest switches to be traversed.

The system then secures the road by switching the switches and setting the signals for the road, marks all tracks and switches on the route as occupied, and updates the track display in the screen window.

The driving order is issued to the train control by setting the corresponding signal to *drive* or *slow drive position*.

When a train has arrived on an entry track from the neighbouring station, it must be "*picked up*" by the train dispatcher, i.e., a driving order (from the entry track as the start track) to a destination track in the station must be issued.

If an exit track is the destination of a journey, the train disappears in the direction of the following station.

In the past, all this was done with mechanical lever interlockings, wire rope pulls, and electromechanical relay controls; nowadays, it is probably done everywhere by computer-controlled sensors and motors.

14.1.5 Representation of Train Journeys

The duration of trains on their journey over tracks of different lengths is greatly simplified: We assume that they travel at a constant speed, i.e., that the duration of a journey over a block depends only linearly on its length.

The journey of a train, its arrival at its destination track, or the fact that it has left the station is visually displayed on the track display. Tracks and switches that the train has left are immediately released on the track display.

14.1.5.1 Inclusion of Multiple Stations

The system consists of several stations.

A distributed solution is planned for this: Each station is operated on its own computer or in a heavyweight process on one of the participating computers; the operation in between is synchronized by the respective neighbouring stations. By "synchronization" we understand that a train appears on the entry track of the destination station after leaving the exit track on a journey to a neighbouring station.

14.2 System Architecture

14.2.1 The Objects of the System

From the system analysis of the project, the following objects can be derived in the planned system, each of which is packed into a package:

- *Stations* as *track diagrams* of the entirety of all blocks.
- The *network* of all involved stations with which the train traffic between the stations is *synchronized*.
- *Roads* as ordered sequences of pairwise connected blocks, which are constructed, occupied, switched, driven, and then released.
- *Blocks* with
    - their *connections* to the neighbouring blocks;
    - their *type (track, switch, or double crossover switch)*;
    - their *location*;
    - their *position on the screen*;
    - the *cells*, they consist of;
    - if they are switches or double crossover switches:
        their *branching direction*, (with or against the kilometre marking),
        their *switch direction* (left switch or right switch), and
        position (left, straight, or right);

- their *signals*; and
- their *state* (free or occupied with a stationary or moving train);
- for each block its *signals*;
- *cells* in the form of a straight or bent track piece, a buffer stop, or a switch and double crossover switch in their respective position, from which the *representations* of the blocks on the screen are composed.

In addition, the system requires the following small packages:

- *kilo* for the both directions ("with" and "against" kilometerage);
- *colour* for the colours needed by the system;
- *direction* for the position of the switches (left, straight, right); and
- various *constants* for positioning on the screen windows.

In the following, we explain these basic components.

14.2.2 Component Hierarchy

In Fig. 14.1, you can find the dependencies of the individual packages on each other: the package that is lower is imported (used) by the one above it. The representation slightly exaggerates the import relationships insofar as

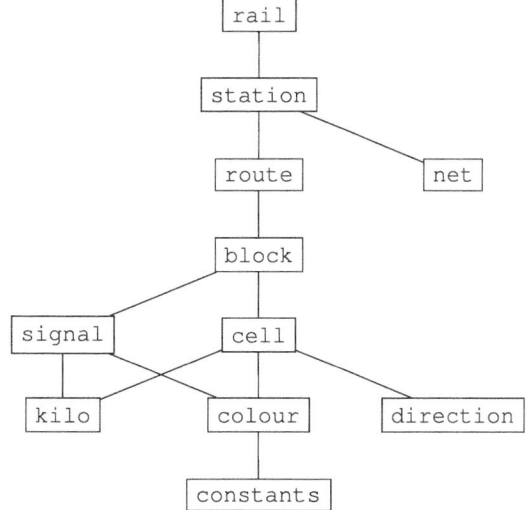

Fig. 14.1 Architecture of Rail

- between the definition and implementation parts of the packages no distinction is made and
- not every import becomes visible, because at some points there are also imports over several layers (for example, the colours are imported by all indirect above modules).

14.3 User Manual

14.3.1 Screen Design

The only alphanumeric formats in the system are the

- Station names (strings of limited length) and
- the track numbers (natural numbers < 100).

14.3.1.1 Representation of the Cells

The cells have a size of 36 pixels horizontally and 24 pixels vertically.

There are cells for the representation of

- tracks,
- bends,
- switches,
- double crossing switches, and
- buffer stops.

Figure 14.2 shows track cells, Fig. 14.3 shows track bends, Fig. 14.4 shows switches branched in the direction of the kilometerage, Fig. 14.5 shows switches that are branched against the direction of the kilometerage, Fig. 14.6 shows double crossover switches, and Fig. 14.7 shows buffer stops.

Fig. 14.2 Track cells

Fig. 14.3 Track bends

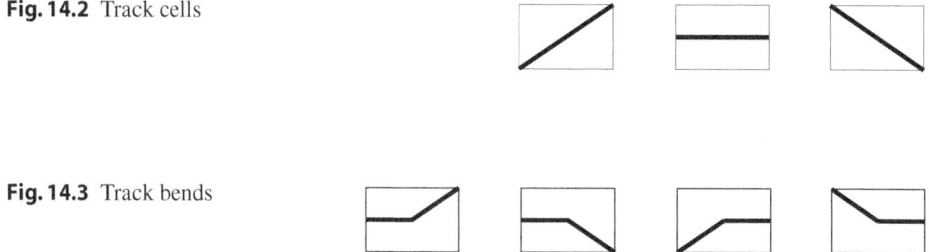

14.3 User Manual

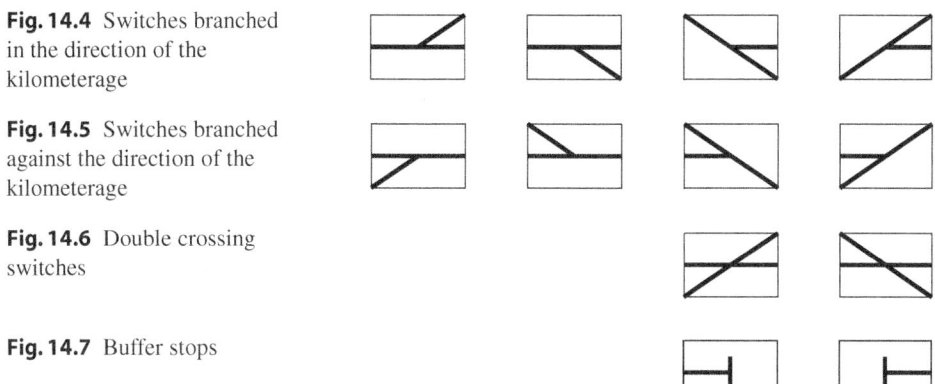

Fig. 14.4 Switches branched in the direction of the kilometerage

Fig. 14.5 Switches branched against the direction of the kilometerage

Fig. 14.6 Double crossing switches

Fig. 14.7 Buffer stops

14.3.1.2 Representation of Blocks

A block is represented as sequences of cells, with track numbers approximately in the middle of the track block for tracks.

Switches and double crossover switches are blocks of length 1, so they are represented as the corresponding cells. Switch numbers are not written, as they do not need to be identified on the track diagram because they are not switched by the train control, but by the system.

14.3.1.3 Representation of Signals

Signals are represented as small (colour-filled) circles; in the direction of the kilometerage at the end of a track block *below* a track cell and in the opposite direction *above* a track cell.

14.3.1.4 Occupancy Reports

The different states of tracks, switches, and routes are displayed as follows:

Free tracks are shown in *green*, tracks occupied by a stationary train are shown in *yellow*, and tracks occupied by a moving train are shown in *red*.

The positions of switches and double crossover switches are evident in that only the switched branches have the corresponding colour, the others are grey.

14.3.2 The Track Diagram Control Panel on the Screen

Input and output are done on a static screen; techniques such as "screen scrolling" are avoided; the train dispatcher must always have the *entire* station in view (there are also no "scrolling pushbutton control panels").

The screen windows have a size of 42 cells vertically and 8 cells horizontally and 4 lines for the distance to the top and bottom edge of the windows and the last screen line for error messages and operating instructions.

This specification limits the number of tracks in the stations: A maximum of eight parallel tracks can be displayed.

After successful execution of the driving order, the route is designated as a road, i.e., marked as occupied by a moving train.

The representation of a train journey in its movement is achieved by cleared track sections and switches immediately assuming the state of being *free*, i.e., changing to the free colour.

14.3.3 The Net of the Stations

Figure 14.8 shows the network of the six stations.

In the following, we present the individual stations.

14.3.4 The Network of Stations

Bahnheim

Bahnheim is a terminal station of a double-track line with two main tracks (2 and 3), two secondary tracks (1 and 4), four storage tracks (11 to 14), and one flank protection track (21) (see Fig. 14.9). A platform is to be imagined between tracks 2 and 3.

Track 23 is the entrance track and track 24 is the exit track to and from Bahnhausen.

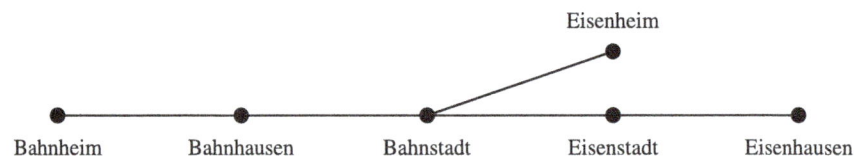

Fig. 14.8 The net of the six stations

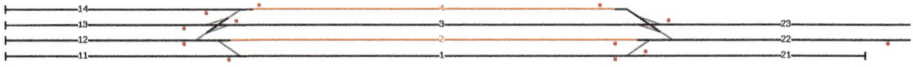

Fig. 14.9 Track diagram of Bahnheim

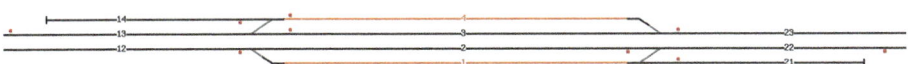

Fig. 14.10 Track diagram of Bahnhausen

Bahnhausen

Bahnhausen is a station on a double-track line with two main tracks (2 and 3), two secondary tracks (1 and 4), and two flank protection tracks (14 and 21) (see Fig. 14.10). A platform is to be imagined between tracks 2 and 3.

Tracks 12 and 13 are the entrance and exit tracks to and from Bahnheim, and tracks 23 and 22 are the entrance and exit tracks to and from Bahnstadt.

Bahnstadt

Bahnstadt is a station on a double-track main line with two main tracks (3 and 4), with sidings (1, 2, and 6), and two flank protection tracks (16 and 22). Track 5 is the starting point for the branching single-track branch line to Eisenheim; tracks 6 to 8 with their sidings form a small freight loading section with pull-out and storage tracks for shunting purposes (see Fig. 14.11).

Platforms are to be imagined between tracks 2 and 3 and between tracks 4 and 5.

Tracks 13 and 34 are the entry tracks from Bahnhausen or Eisenstadt and tracks 14 and 33 are the corresponding exit tracks. Track 36 is the entry and exit track from or to Eisenheim.

Eisenheim

Eisenheim is a terminus of a single-track branch line with one main track (2), two sidings (1 and 3), a flank protection track (13), and three storage tracks (21 to 23) (see Fig. 14.12). A platform is to be imagined between tracks 2 and 3.

Track 12 is the entry and exit track from or to Bahnstadt.

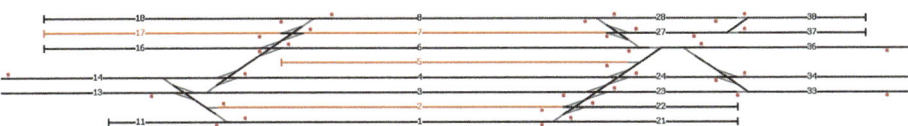

Fig. 14.11 Track diagram of Bahnstadt

Fig. 14.12 Track diagram of Eisenstadt

Eisenstadt

Eisenstadt is a through station on a double-track main line with two main tracks (2 and 3), two sidings (1 and 4), and two flank protection tracks (14 and 21) (see Fig. 14.13). A platform is to be imagined between tracks 2 and 3.

Tracks 12 and 23 are the entry tracks from Bahnstadt or Eisenhausen and tracks 13 and 22 are the corresponding exit tracks.

Eisenhausen

Eisenhausen is a terminus on a double-track main line with two main tracks (2 and 3), two sidings (1 and 4), a flank protection track (14), and four storage tracks (21 to 24) (see Fig. 14.13. As with the previous stations, a platform is to be imagined between tracks 2 and 3.

Tracks 12 and 13 are the entry and exit tracks from or to Eisenstadt.

14.3.5 System Operation

It is very simple. After starting the server, operations can be started at the stations, which consist of driving orders.

14.3.5.1 Station Selection

Each station is controlled by its own computer or a heavyweight process on a computer.

The system is responsible for synchronizing train traffic between stations. For this, it is necessary to start the server—a component in the network package—before calling up operations at one of the stations. This is done by calling train (without arguments).

Afterwards, a station is selected by calling "train n", where n is one of the numbers from 0 to 5: It is then the n-th station from the network (in the order specified in Sect. 14.4.2).

Fig. 14.13 Track diagram of Eisenhausen

The screen window shows its track image in the state defined in the station package and the station's operation can be started.

14.3.5.2 Issuing a Driving Order
The start and destination tracks for a route are selected as follows:

There is the hint "Click start track end operation: Esc". By clicking on the corresponding track with the left mouse button, the start track is selected; the escape key ensures that operations at this station—but only after all train journeys have been completed—are stopped.

Then there is the hint "Click destination track other start track: Esc"; here too, the destination track is clicked with the left mouse button.

If a *different* start track is to be selected, the escape key must be pressed.

The following errors are possible with these clicks:

- As a *start track*, a *free track section* or
- an *exit track* of a double-track connection or
- as a *destination track*, an *occupied track section* or
- an *entry track* of a double-track connection or
- as a *start track* or *destination track*, a *switch* is specified.
- there are no routes from the start to the destination track or
- there are routes, but all of them have some tracks occupied by standing trains or train journeys.

In the first five cases, the driving order is simply ignored by the system; in the other two cases, an appropriate error message is issued.

After selecting the start and destination tracks, the system searches for the route with the fewest switches among the possible routes.

After the successful issuance of the driving order, the switches and signals are set, and then the train departs.

14.4 Construction

We only show the specifications of the library packages and the representations of the respective abstract data objects.

14.4.1 Main Program

The main program is very short, it consists only of the activation of the server or the start of operations at one of the stations.

```
package main
import ("µU/ker"; "µU/env"; "µU/kbd"; "rail/station"; "rail/net")

func main() {
  if env.NArgs() == 0 {
    net.MyStation = net.Server
    net.Activate()
    kbd.Wait (false)
    return
  }
  n, m := env.N(1), net.N - 1
  if n >= m {
    ker.Panic ("the argument of the call \"rail\" has to be smaller than ", m)
  }
  net.MyStation = n
  station.New().Operate()
}
```

14.4.2 Network

The tasks of the network package are

- the management of the six stations with their connections to each other and
- the work of the server, which occupies and releases the entrances to the stations.

Its network specification reads

```
package net

const M = 11 // maximal length of the names of stations
const (Bahnheim = uint(iota); Bahnhausen; Bahnstadt; Eisenheim;
       Eisenstadt; Eisenhausen; Server; N; A = N - 1)
var (MyStation uint; MyName string)

func Name (n uint) { return name() }

func NumNeighbours() uint { return numNeighbours() }

func Neighbour (n, i uint) uint { return neighbour (n,i) }

func Activate() { activate() }

func ClearEntrance() { clearEntrance() }

func OccupyEntrance() { occupyEntrance() }
```

The implementation of the network package includes two files; in network.go the names of the stations and their respective neighbours are defined and in monitor.go the abstract data type mon and the computer on which the server runs are defined:

14.4 Construction

```
package net
import (. "µU/obj"; "µU/host"; "µU/fmon")

const ( clear = uint(iota); occupy; occupied; nOps)

type Entrance interface {

  ClearEntrance (n uint)
  OccupyEntrance (n uint)
  EntranceOccupied (n uint) bool
}
type mon struct {
               fmon.FarMonitor
               }
var (
  server = host.Localhost().String()
  monitor Entrance
  aktiv = false
  entryFree [A*A]bool
)

func init() {
  for n := uint(0); n < A; n++ {
    for i := uint(0); i < NumNeighbours (n); i++ {
      entryFree[A * n + Neighbour (n, i)] = true
    }
  }
}

func activate() {
  monitor = New (server, 2345, MyStation >= N - 1)
}

func New (h string, p uint16, s bool) Entrance {
  fs := func (a Any, i uint) Any {
          n := a.(uint)
          switch i {
          case clear:
            entryFree[n] = true
          case occupy:
            entryFree[n] = false
          case occupied:
            if ! entryFree[n] { return uint(1) }
          }
          return uint(0)
        }
  m := new(mon)
  m.FarMonitor = fmon.New (uint(0), nOps, fs, AllTrueSp, h, p, s)
  return m
}

func (m *mon) ClearEntrance (n uint) {
  m.F (n, clear)
}

func (m *mon) OccupyEntrance (n uint) {
  m.F (n, occupy)
}

func (m *mon) EntranceOccupied (n uint) bool {
  return m.F (n, occupied).(uint) == 1
}

func clearEntrance (n uint) {
  monitor.ClearEntrance (n)
}
```

```
func occupyEntrance (n uint) {
  monitor.OccupyEntrance (n)
}

func entranceOccupied (n uint) bool {
  return monitor.EntranceOccupied (n)
}
```

14.4.3 Stations

The only task of the stations is their operation. Therefore, the specification of the station package is very short:

```
package station

type Station interface {
  Operate()
}

func New() Station { return new_() }
```

The representation of the abstract data type Station is very simple, it consists only of the graph of the blocks, represented by their numbers, and the indication of the kilometerage, which always indicates the direction of the current route:

14.4.4 Routes

Routes and route sequences are ordered sequences of blocks. They have a start block and a target block and blocks can be arranged in them.

```
package route

type Route interface {

// Returns the number of the start block of x.
  Start() uint

// Returns the number of the destination block of x.
  Destination() uint

// x does not contain any blocks.
  Clr()

// The block with the number n is inserted in order.
  Insert (n uint)

// Returns the number of the i-th block of x.
  Nr (i uint) uint

// Returns the number of blocks in x.
  Num() uint

// Returns true, iff the i-th block in x
// is smaller than the j-th block.
  Less (i, j int) bool
```

```
// Returns true, if the i-th block in x
// is smaller than the j-th block or if it is the same.
  Leq (i, j int) bool

// Returns true, if x contains switches or double crossing switches
// with a deflecting position (left or right).
  Deflecting() bool
}

// Returns a new empty route,
func New() Route { return new_() }
```

14.4.5 Blocks

The blocks are the basic components of a station, because every station is a graph whose nodes are the blocks. Each block consists of an ordered sequence of cells (see Sect. 14.4.6). There are

- straight track blocks,
- bends (bent track blocks),
- switches, and
- double crossing switches (blocks).

The last three ones consist only of one cell.

Blocks are numbered; in the case of track blocks, these are the track numbers. There are different sorts of tracks:

- *Through tracks*,
- *Entry and Exit tracks*, and
- *Siding tracks*.

Each block has one of the states

- free, i.e., not occupied by a train,
- occupied, i.e., occupied by a stationary train, and
- travelling, i.e., occupied by a moving train.

Here is the specification of the block package:

```
package block
import (. "µU/obj"; "µU/col";
        . "rail/kilo"; . "rail/direction"; "rail/signal")

const H = 100
type Kind byte; const (
```

```
   Dfg = Kind(iota) // ThroughTrack
   AsM // SidingWith
   AsG // SidingAgainst
   EfM // EntryTrackWith
   EfG // EntryTrackAgainst
   AfM // ExitTrackWith
   AfG // ThroughTrackAgains
   EAM // EntryExitTrackWith
   EAG // EntryExitTrackAgainst
   Bend
   Switch
   DCS // Double crossing switch
   NKinds
)
type Block interface {

   Object
   Stringer

// Pre: x is not empty.
// Returns the number of x modulo M.
   Numbershort() uint

// Returns the number of x.
   Number() uint

// Returns the inclination of x.
   Inclination() Direction

// Pre: z < NLines, s + 1 <= NColumns, s < sn < s + 1.
//      The position (z, s) is not yet occupied by a block.
//      x starts at (z, s) and runs straight to the right
//      (a = straight/left/right: horizontally/diagonally ascending/
//      diagonally descending) with the column length l.
// x has the number n, it is displayed in the column sn. x has the signal
// of type gt in direction g with the position gst at (gz, gsn) and that
// of type mt in direction m with the position mst at (mz, msn).
   DefineTrack (n uint, k Kind, d Direction, l, z, s, sn uint,
                gt signal.Typ, g Kilometerage, gst signal.Position, gz, gsn uint,
                mt signal.Typ, m Kilometerage, mst signal.Position, mz, msn uint)

// Returns true, iff x is a track.
   IsTrack() bool

// Returns true, iff x is a through track.
   IsThroughtrack() bool

// Returns true, iff x is an entry track.
   IsEntrytrack() bool

// Returns true, iff x is an exit track.
   IsExittrack() bool

// Returns true, iff x is an entry exit track.
   IsEntryExittrack() bool

// Returns true, iff x is a siding track in direction k.
   IsSidingtrack (k Kilometerage) bool

// x is a bend with the number n in direction k
// with the buckling direction r at (z, s).
   DefineBend (n uint, k Kilometerage, r Direction, z, s uint)

// Returs true, iff x is a bend.
   IsBend() bool

// Pre: z < NLines, s < NColumns, l != r, r = Left or Right.
```

14.4 Construction

```
//      The position (z, s) is not yet occupied by a block.
// x is not empty. x has the mumber n.
// k is the kilometerage, in which the switch is branched.
// l is the position of the continuous branch of the switch.
// (l = Straight/Left/Right: horizontally/diagonally ascending/descending).
// x is for r == Left resp. Right a left resp. right switch
// with the position st at (z, s).
   DefineSwitch (n uint, k Kilometerage, l, r, st Direction, z, s uint)

// Returns true, iff x is a switch.
   IsSwitch() bool

// Pre: l != Straight.
// x is a DCS with the number n, the inclination l and the position (z, s).
   DefineDCS (n uint, l, r Direction, z, s uint)

// Returns true, iff x is a DCS.
   IsDCS() bool

// Pre: x is a switch.
// Returns the direction of the branch x.
   SwitchDirection() Direction

// Pre: x is a switch.
// Returns the direction of the kilometerage, in which x is branched.
   BranchingDirection() Kilometerage

// Returns the position at the left edge of x.
   Pos() (uint, uint)

// Returns the position at the left edge of x.
   Line() uint

// Pre: x is a switch or a DCS.
// x has the position r.
   Set (d Direction)

// Pre: x is a switch.
// Returns the position of x.
   Position() Direction

// If x has a signal in direction k, it has the position s.
// The signal is displayed.
   SetSignal (k Kilometerage, s signal.Position)

// x is displayed.
// If x is a track with a number > 0, this number is also displayed.
   Write (f col.Colour)

// Pre: x is not empty. l < NLines, c < NColumns.
// Returns true, iff x occupies the position (l, c).
   PositionOccupied (l, c uint) bool

// x is not occupied.
   Clr()

// Returns true, iff x is not occupied.
   Free() bool

// Pre: x is not empty.
// x is occupied by a standing train.
   Occupy()

// Pre: x is not empty.
// x is occupied by a moving train.
   Travelling()
```

```
// Pre: x is not empty.
// x is occupied by a standing train and blinks.
  OccupyArrival()

// Returns true, iff x is occupied by a train.
  Occupied() bool

// Returns the colour of x depending on the state free, occupied or used.
  Colour() col.Colour

// Pre: x is not empty.
// Returns the type of the signal of x in direction k, if there is one;
// otherwise NT.
  Signaltyp (k Kilometerage) signal.Typ

// x blinks for a moment.
  Blink()

// Returns the column length of x.
  Length() uint

// Returns true, iff the mouse pointer points to x.
  UnderMouse() bool

// Returns true, iff x is a switch or a DCS with the branching direction k.
  Branched (k Kilometerage) bool
}

var
  Nr []uint
const
  M = 300
var
  B, W, D [M]Block

// Returns a new empty block.
func New() Block { return new_() }

// Returns the number of the pairs.
func NPairs() uint { return nPairs() }

// Pre: i < NPairs().
// Returns the i-th pair.
func Pair (i uint) (uint, uint) { return pair(i) }

// Returns the number of the block, iff it is under the mouse;
// in this case it is > 0. Returns otherwise 0.
func Found() uint { return found() }
```

The representation of blocks is somewhat more complex:

```
package block
import ("µU/ker"; . "µU/obj"; "µU/time"; "µU/col"; "µU/scr"; "µU/str"
        "µU/N"; "µU/seq"; . "rail/colour"; . "rail/kilo"; . "rail/direction"
                         . "rail/constants"; s "rail/signal"M; "rail/cell")

type state byte; const (free = state(iota); occupied; travelling)

type block struct { uint32 "number"
                    Kind
                    Kilometerage // branching direction, if switch
             location,
             direction,
               position Direction
                    uint // length = number of cells
```

14.4 Construction

```
              1, c uint // position at the left border
                 seq.Sequence // sequence of the cells
                 state
              sig [NK]s.Signal
}
```

14.4.6 Cells

Cells are the components of blocks. There are the following types of cells:

- tracks,
- bends,
- switches,
- double crossing switches, and
- buffer stops.

Cells have a location, a direction, possibly a position, and a position in the screen window (see Sect. 14.3.1.1).

The specification of the cell package is

```
package cell
import (. "µU/obj"; "µU/col"; . "rail/kilo"; . "rail/direction")

type Cell interface {

  Object

// x has the number n.
  Renumber (n uint)

// Returns the number of x.
  Number () uint

// Pre for all methods with the parameters (z, s) at the end, that
//     define a cell: The position (z, s) is not yet occupied by a cell.

// x is a track with the inclination a and the position (z, s) on the screen.
  Track (n uint, a Direction, z, s uint)

// Returns true, if x is a track.
  IsTrack () bool

// x is a bend with nunber n in the direction of the kilometerage k
// at position (z, s) bended at direction d.
  Bend (n uint, k Kilometerage, d Direction, z, s uint)

// x is a buffer stop in direction k at position (z, s).
  BufferStop (k Kilometerage, z, s uint)

// x is for r = right a right switch, otherwise a left switch
// with number n, branching direction k, inclination l,
// position st and screen positioo (z, s).
  Switch (n uint, k Kilometerage, l, r, st Direction, z, s uint)

// Returns (k, true), iff x is a switch with the branching direction k.
```

```
  IsSwitch() (Kilometerage, bool)

// Pre: l != Straight.
// x is a double crossing switch with number n, inclinatin l, position r
// and position (z, s) on the screen.
  DCS (n uint, l, r Direction, z, s uint)

// Returns (k, true), iff x is a double crossing switch with
// with branching direction k.
  IsDCS() (Kilometerage, bool)

  String() string

// If x is a switch or a double crossing switch,
// it is set in direction r.
  Set (r Direction)

// Returns the kilometerage of x.
  Kilo() Kilometerage

// Returns the inclination    of x in direction k (if x is branched
// in direction k the inclination of the continuous branch)
// and the position of x on the screen.
  Inclination (k Kilometerage) (Direction, uint, uint)

// Returns the position of x, if x is a switch or a double crossing switch
// returns otherwise Straight.
  Position() Direction

// Returns true, iff x has the position (z, s) on the screen.
  HasPosition (z, s uint) bool

// Returns the position of x on the screen.
  Pos() (uint, uint)

// x is written to the screen in colour c.
  Write (c col.Colour)

// Returns true, iff the mouse pointer points to x.
  UnderMouse() bool
}

func New() Cell { return new_() }
```

The representation of the cells is also somewhat complex:

```
package cell
import ("µU/ker"; . "µU/obj"; "µU/col"; "µU/scr"; "µU/N"; . "rail/kilo"
  . "rail/direction"; . "rail/colour"; . "rail/constants";)

const max = 15

type kind byte
const (track = kind(iota); bend; sw // switch
       dcs; bufferstop; nk)
type cell struct { kind
                   uint32 "number"
                   Kilometerage
         location,
         direction,
           position Direction
             z, s uint // position on the screen
         lastColour col.Colour
}
```

14.4.7 Signals

Signals are components of blocks: Each block can have one or two signals, one in the direction of the kilometre marking and/or one in the opposite direction.

We only include main signals in the system. They always have one of the following positions:

- Hp0 = Stop,
- Hp1 = Go, or
- Hp2 = Slow movement.

The specification of the signal package is quite short:

```
package signal
import (. "µU/obj"; . "rail/kilo")

type Typ byte; const (
  T0 = Typ(iota);
  T1 // Hp0, Hp1
  T2 // Hp0, Hp1, Hp2
      NT
)
type Position byte; const (
  Hp0 = Position(iota) // stop
  Hp1 // Fahrt
  Hp2 // Langsamfahrt
  NS
)
type Signal interface {

  Object

// x is defined, has the nunber n, the type t, the kilometerage k,
// the position st and the position (z, s) on the screen.
  Define (n uint, t Typ, k Kilometerage, st Position, z, s uint)

// Returns the Typ of x.
  Signaltyp() Typ

// Pre: x is defined.
// x has the Position s and is written to the screen at its screen position.
  Set (s Position)

// If x is defined, it is written to the screen at its screen position.
  Write()
}

func New() Signal { return new_() }
```

14.4.8 Aid Packages

In addition to the packages presented so far, a few small packages are still needed.

14.4.8.1 Colour

The different states of the blocks are distinguished by colours:

- green for free,
- yellow for occupied, and
- red for travelling.

This results in the specification of the colour package:

```
package colour
import "µU/col"

var (Foregroundcolour, Backgroundcolour, Nocolour, Freecolour,
  Occupiedcolour, Railcolour, Travelcolour, Slowtravelcolour,
  Stopcolour col.Colour)
```

14.4.8.2 Kilometerage

The kilometerages are
 • With = in the direction of the kilometre marking and • Against = against this direction.
The specification of the kilometre marking package is trivial:

```
package kilo

type Kilometerage byte; const (With = Kilometerage(iota); Against; NK)

var Ktext = [NK + 1]string ("With", "Against", "NK")

// Returns Against for k == With, otherwise With.
  func OppositeDirection (k Kilometerage) Kilometerage
  { return opposite(k) }
```

14.4.8.3 Directions

The directions

- left,
- straight, and
- right

are used for the inclination of cells and the positions of switches. The specification of the direction package is also very simple:

```
package direction

type Direction byte; const (
  Left = Direction(iota); Straight; Right; ND)

var Dtext = [ND+1]string ("Left", "Straight", "Right", "ND")

func Opposite (d Direction) Direction { return opposite(d) }
```

The size and positioning of cells (thus of blocks) on the screen windows serve various constants, which can be found in the corresponding package:

```
package constants

var Y0, H1, H2, W1, W2 int
const (NLines, NColumns = 16, 42)

func Init() { init_() }
```

14.4.9 Other Packages

At deeper levels, many other packages are needed, which are components of the microuniverse due to their universal usability, such as sequences (μU), persistent sequences (μU), graphs (μU), and my egg-laying wool milk pig for distributed problems, the remote monitor (μU).

References

1. Eisenbahn-Bau- und Betriebsordnung. https://www.gesetze-im-internet.de/ebo/
2. Richtlinie 408 – Fahrdienstvorschrift der Deutschen Bahn
3. Eisenbahn-Signalordnung 1959. https://www.gesetze-im-internet.de/eso_1959/

Figures in Space

15

So you look, huh?

Abstract

This chapter is about the spatial representation of classical figures.

In my lectures in teacher training for mathematics, I occasionally drew sketches of structures from analytical geometry in three-dimensional space on the board to illustrate various concepts and to connect with prior knowledge in the context of "modules and vector spaces".

As a byproduct of some graphic packages from the microuniverse, the package `rfig` was created for visualizing simple scenarios from analytical geometry, with which, for example, conic sections can be vividly represented.

This project is simply about generating and being able to view standard figures in three-dimensional space.

The advantage over a (two-dimensional) sketch is obvious; for example, being able to "walk around" the intersections of (double) cones with planes in 3D space—even if only virtually—is considerably more illustrative than lousy 2D sketches …

15.1 System Analysis

It is intended to represent *three-dimensional figures in space*, both angular ones, such as cubes, pyramids, or octahedrons, and round ones, such as spheres, cones, cylinders, and tori.

15.1.1 System Architecture

15.1.2 The Objects of the System

These are the three-dimensional figures from the package μU:

- *points*;
- *lines* and *sequences of lines*;
- *triangles* and *sequences of triangles*;
- *quadrilaterals* and *sequences of quadrilaterals*;
- horizontal and vertical *rectangles*;
- *parallelograms*;
- *polygons*;
- *curves* given by functions;
- *planes*, given by function terms of the form f (x, y) = ax + by + c;
- *cubes*;
- *cuboids* with horizontal base;
- general *prisms*;
- *parallelepipeds*;
- *pyramids* and *multipyramids*;
- *octahedra*;
- horizontal and vertical *circles* and *segments of circles*;
- *spheres*, given by the coordinates of the centre (x, y, z) and their radius r;
- *cones* and *double cones* (with vertical axis of symmetry), given by the coordinates of their apex (x, y, z), their radius f and their height h;
- *cylinders*, *segments of cylinders*, horizontal *cylinders*;
- horizontal and vertical *toruses*;
- *paraboloids*; and
- *surfaces* given by functions.

15.2 Component Hierarchy

There is only the main program `spacefig.go`, which uses the package `fig3`, therefore the hierarchy is very flat (see Fig. 15.1).

Fig. 15.1 Component hierarchy of the spatial figures

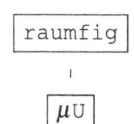

15.3 User Manual

The use of the system consists in writing similar short programs as shown above, which requires the study of the specifications of the OpenGL package μU.

15.4 Construction

15.4.1 Specifications

For the specification, that of the OpenGL package μU is important; the implementations are to be written as very short programs in the style of those presented in Sect. 15.4.3.

The three-dimensional figures

The specification of μU contains many figures. However, we only show the section that deals with three-dimensional figures here, and also leave out the functions in which figures with multiple colours occur:

```
package fig3
import (. "µU/obj"; "µU/col")

// The specifications of all functions
// are found in the file µU/gl/def.go.
func Cube (c col.Colour, x, y, z, a float64) { cube(c,x,y,z,a) }
func Cuboid (c col.Colour, x, y, z, x1, y1, z1 float64) {
              cuboid(c,x,y,z,x1,y1,z1) }
func Prism (c col.Colour, x ...float64) { prism (c,x...) }
func Parallelepiped (c col.Colour, x ...float64) {
                    parallelepiped(c,x...) }
func Pyramid (c col.Colour, x, y, z, a, h float64) {
             pyramid(c,x,y,z,a,h) }
func Multipyramid (f col.Colour, x, y, z, h float64, c ...float64) {
                  multipyramid(f,x,y,z,h,c...)}
func Octopus (c col.Colour, x ...float64) { octopus(c,x...) }
func Octahedron (c col.Colour, x, y, z, r float64) {
                octahedron(c,x,y,z,r) }
func OctahedronC (c []col.Colour, x, y, z, r float64) {
                 octahedronC(c,x,y,z,r) }
func Sphere (c col.Colour, x, y, z, r float64) { sphere(c,x,y,z,r) }
func Cone (c col.Colour, x, y, z, r, h float64) { cone(c,x,y,z,r,h) }
func DoubleCone (c col.Colour, x, y, z, r, h float64) {
                doubleCone(c,x,y,z,r,h) }
func Cylinder (c col.Colour, x, y, z, r, h float64) {
              cylinder(c,x,y,z,r,h) }
func CylinderSegment (c col.Colour, x, y, z, r, h, a, b float64) {
                     cylinderSegment (c,x,y,z,r,h,a,b) }
func HorCylinder (c col.Colour, x, y, z, r, l, a float64) {
                 horCylinder(c,x,y,z,r,l,a) }
func Torus (c col.Colour, x, y, z, R, r float64) {
           torus(c,x,y,z,R,r) }
func VerTorus (c col.Colour, x, y, z, R, r, a float64) {
              verTorus(c,x,y,z,R,r,a) }
func Paraboloid (c col.Colour, x, y, z, a, wx, wy float64) {
                paraboloid(c,x,y,z,a,wx,wy) }
func Surface (c col.Colour, f Fxy2z, wx, wy float64) {
             surface(c,f,wx,wy) }
```

15.4.2 Implementations

The implementation of μU consists only of direct accesses to the OpenGL package μU from the microuniverse. To this end, we show a section from the specification of this package:

```
// Pre: wx > 0, wy > 0.
// The bounded plane within the area -wx <= x <= wx and -wy <= y <= wy,
// defined by f(x,y) = a * x + b * y + c, is created.
func Plane (a, b, c, wx, wy float64) { plane(a,b,c,wx,wy) }

// Pre: a != 0.
// A cube with edges parallel to the coordinate axes is created
// with the center at (x, y, z) and the edge length a.
func Cube (x, y, z, a float64) { cube(x,y,z,a) }

// Pre: len(x) == 6, x[0] != x[3], x[1] != x[4] and x[2] != x[5].
// A cuboid with edges parallel to the coordinate axes is created
// between the points at (x[0], x[1], x[2]) and (x[3], x[4], x[5]).
func Cuboid (x ...float64) { cuboid (x...) }

// Pre: len(x)
// A prism without bottom and top is created.
// Its bottom corners are (x[3], x[4], x[5]), (x[6], x[7], x[8])
and so on,
// its top corners are the bottom corners plus (x[0], x[1], x[2]).
func Prism (x ...float64) { prism (x...) }

// Pre: len(x) == 12.
// A parallelepiped is created.
// One of its corners is c = (x[0], x[1], x[2]), the others are
// c + (x[3], x[4], x[5]), c + (x[6], x[7], x[8]) and c + (x[9],
x[10], x[11]).
func Parallelepiped (x ...float64) { parallelepiped (x...) }

// Pre: a > 0, h != 0.
// A pyramid of height h with the center (x, y, z) of its horizontal
bottom
// is created, its bottom edges have the length a.
func Pyramid (x, y, z, a, h float64) { pyramid (x,y,z,a,h) }

// Pre: len(x)
// An octopus with top (x[0], x[1], x[2]) and corners (x[3], x[4],
x[5]),
// (x[6], x[7], x[8]) and so on is created.
func Octopus (x ...float64) { octopus (x...) }

// Pre: r != 0.
// An octahedron with the center (x, y, z) and length e of its edges is
created.
func Octahedron (x, y, z, e float64) { octahedron (x,y,z,e) }

// Pre: r != 0.
// A sphere is created with the center (x, y, z) and the radius r.
func Sphere (x, y, z, r float64) { sphere (x,y,z,r) }

// Pre: r != 0, h != 0.
// A cone of height h is created with the horizontal circle
// around (x, y, z) with radius r as its bottom.
func Cone (x, y, z, r, h float64) { cone (x,y,z,r,h) }

// Pre: r != 0, h != 0.
// Two cones of height h are created, one with the horizontal circle
// around (x, y, z - h) as bottom and the other with the horizontal
circle
// around (x, y, z + h) as top.
```

15.4 Construction

```
func DoubleCone (x, y, z, r, h float64) { doubleCone (x,y,z,r,h) }

// Pre: r != 0, h != 0.
// A cylinder of radius r and height h is created with the
// horizontal circle around (x, y, z) with radius r as bottom and
// the horizontal circle around (x, y, z + h) as top.
func Cylinder (x, y, z, r, h float64) { cylinder (x,y,z,r,h) }

// Pre: R > 0, r > 0.
// A horizontal torus with the center at (x, y, z),
// the inner radius R-r and the outer radius R+r is created.
func Torus (x, y, z, R, r float64) { torus (x,y,z,R,r) }

// Pre: a != 0, wx > 0, wy > 0.
// A paraboloid within the area -wx <= x <= wx and -wy <= y <= wy
is created
// with base point (x0, y0, z0), defined by
// f(x, y) = a$^2$ * ((x - x0)$^2$ + (y - y0)$^2$).
func Paraboloid (x0, y0, z0, a, wx, wy float64) { paraboloid(x0,y0,z0,
a,wx,wy) }

// Pre: wx > 0, wy > 0.
// The bounded surface within the area -wx <= x <= wx and -wy <= y
<= wy,
// given by the function f is created.
func Surface (f obj.Fxy2z, wx, wy float64) { surface (f,wx,wy) }
}
```

The main program relies on the function Go from the package μU for the representation of 3D scenes. We briefly explained this in Chap. 13 about Lindenmayer systems (see Sect. 13.3.6).

15.4.3 Examples

The program
```
package main
import ("µU/col"; "µU/gl"; "µU/scr"; . "µU/fig3")

func main() {
  s := scr.NewWH (0, 0, 800, 600); defer s.Fin()
  gl.ClearColour (col.FlashWhite());
  s.Go (draw, 3,-1, 10, 3,-1, 0, 0, 1, 0)
}

func draw() {
  r, o, y, g := col.Red(), col.Orange(), col.Yellow(), col.Green()
  c, b, m, n := col.Cyan(), col.Blue(), col.Magenta(), col.Brown()
  MultipyramidC ([]col.Colour {m, n, r, o, y, g, c, b},
    0, 2, 0, 2, 3, 3, 1, 4, -1, 4, -2, 3, -2, 2, -1, 0, 2, 0)
  OctahedronC ([]col.Colour{r,o,y,g,c,b,m,n}, 6, 2, 0, 1.4)
  PrismC ([]col.Colour {c, b, m, r, o, y, g}, 1, 0, 1.5, 1, -2, 0,
    -1, -1, 0, -2, -2, 0, 0, -3, 0, -1, -4, 0, 0, -5, 0, 2, -4, 0)
  ParallelepipedC ([]col.Colour{r,o,y,g,b,m}, 5, -2, 0,
    2, 0, 1, 1, -2, 0, -1.5, 0, 2)
}
```
shows a multipyramid, an octahedron, a prism, and a parallelepiped (see Figs. 15.2 and 15.3).

Fig. 15.2 Several figures

Fig. 15.3 Another view of the several figures

In the program
```
package main
import ("µU/col"; "µU/scr"; . "µU/fig3")

func main() {
  s := scr.NewWH (0, 0, 800, 600); defer s.Fin()
  gl.ClearColour (col.FlashWhite());
  s.Go (draw, 0, -12, 0, 0, 0, 0, 0, 0, 1)
}

func draw() {
  m, o := col.Magenta(), col.Orange()
  Sphere (col.Red(), -1.25, -0.5, 0, 2)
```

15.4 Construction

Fig. 15.4 Sphere, tori, and cylinder

```
  Torus (col.Green(), 0, 0, 0, 5, 1)
  VerTorus (col.Blue(), 5, -2, 0, 3, 0.5, 65)
  CylinderC ([]col.Colour{m,o}, 2.0, 1.5, -4, 1, 8)
}
```

a sphere, two tori, and a cylinder are modelled (see Fig. 15.4).

15.4.4 Examples of Conic Sections

These are the figures that were the reason for me to construct this project.

15.4.4.1 Circles and Ellipses

These are the figures that were the reason for me to construct this project.

If you cut a cone with a plane that is horizontal to the cone axis, a circle results as the intersection. If the plane is not horizontal to the cone axis and its angle of inclination is smaller than the angle of inclination of the cone, the intersection is an ellipse. This second case is modelled by the following example program:

```
package main
import ("µU/col"; "µU/gl"; "µU/scr"; . "µU/fig3")
func main() {
  s := scr.NewWH (0, 0, 800, 600); defer s.Fin()
  gl.ClearColour (col.FlashWhite()); gl.Clear()
  s.Go (draw, 0, -6, 2, 0, 0, 2, 0, 0, 1)
}

func draw() {
  Cone (col.Blue(), 0, 0, 0, 2, 5)
  Plane (col.Orange(), 0.8, 0.8, 2.5, 2.5, 2.5)
}
```

Fig. 15.5 Section of a cone with a plane

Fig. 15.6 Another view of this section

Figures 15.5 and 15.6 illustrate this example.

15.4.4.2 Parabolas and Hyperbolas

If you replace the cone with a double cone and the angle of inclination of the plane is greater than the angle of inclination of the cone, the intersection is generally a parabola. The special case of this example, that the plane is parallel to the cone axis, results in a hyperbola as the intersection. This is modelled by the following example program:

```
package main
import ("µU/col"; "µU/gl"; "µU/scr"; . "µU/fig3")
```

15.4 Construction

```
func main() {
  s := scr.NewWH (0, 0, 800, 600); defer s.Fin()
  s.ScrColourB (col.FlashWhite()); scr.Cls()
  s.Go (draw, 5, -3, 2, 0, 0, 0, 0, 1, 0)
}

func draw() {
  gl.ClearColour (col.FlashWhite()); gl.Clear()
  DoubleCone (col.Red(), 0, 0, 0, 1, 3)
  VertRectangle (col.Green(), .3, -1, -3, .3, 1, 3)
}
```

Figures 15.7 and 15.8 show two views of this example.

Fig. 15.7 Section of a double cone with a plane parallel to the cone axis

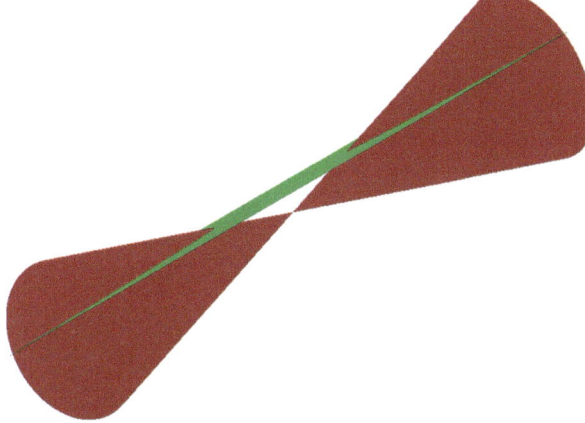

Fig. 15.8 The hyperbola

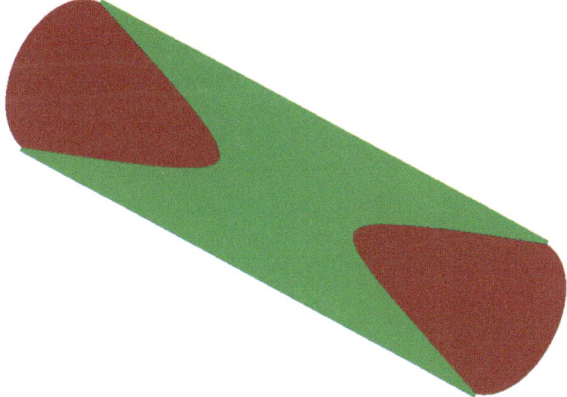

Berlin's U- and S-Bahn

*Berlin, Berlin,
wir fahren durch Berlin.*

inscription on a BVG bus

Abstract

This project shares with the railway project the focus on shortest connections, but here between stations when travelling with Berlin's U- and S-Bahn trains. The network of these trains is displayed on the screen; trips can be found with mouse clicks.

Graph theory also plays a central role in this project: It's about searching for the best connections between two stations in the transport network of Berlin's U- and S-Bahn, a classic example of a graph.

16.1 System Analysis

A *U- or S-Bahn line* consists of *stations* and the *connecting routes* between them. The *network* is the entirety of all U- and S-Bahn lines.

The attributes of the stations are

- their *names*;
- their *coordinates* (latitude and longitude);
- the *lines*, on which they are located;
- an internal *number*, by which they are identified; and
- whether they are a *transfer station* or not.

A *connection* (between two stations) consists of

- the *line*, on which the stations are located;
- the *coordinates* of the two stations it connects; and
- a *natural number* as the average *travel time* between these two stations.

A line *consists* of

- its *designation* (for U-Bahn a "U", for S-Bahn an "S", followed by the *line number* and
- the *colour*, with which they are marked on maps.

16.2 System Architecture

16.3 The Objects of the System

After the system analysis, we have the following objects:

- the network,
- the stations,
- the connections, and
- the lines.

The corresponding packages are the abstract data types

- `net`,
- `station`,
- `track`, and
- `line`.

`net` is an abstract data object, `station` and `track` are abstract data types, and `line` only defines the names and texts of the lines.

16.4 Component Hierarchy

Figure 16.1 shows the dependency of the packages on each other.

Fig. 16.1 Architecture of BUS

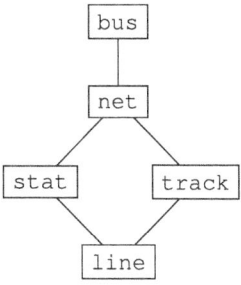

16.5 User Manual

16.6 Construction

Using the program is incredibly simple:

After calling up the program, a graphic appears (see Fig. 16.2), which represents the U- and S-Bahn networks. After clicking on the *start station* and then clicking on the *destination station*, the shortest connection is highlighted in colour.

The graphic can be reduced or enlarged with the enter key ⏎ and the backspace key ⬅, and moved with the arrow keys; the program is terminated with the escape key Esc.

16.6.1 Specifications

Here we show the specifications of the involved packages.

The Network
The specification of the network is very short:

```
package net

// Returns true, iff start- and destination-station were clicked.
func StartAndDestinationSelected() bool { return selected() }

// Shows the shortest connection between the clicked stations.
func ShortestPath () { shortestPath() }
```

The Stations
Stations have the type Object—the prerequisite for them to be inserted as nodes in graphs (see Sect. 3.5.11).

```
package stat
import (. "µU/obj"; . "bus/line")

const (L = 'l'; R = 'r'; O = 'o'; U = 'u')
```

308   16   Berlin's U- and S-Bahn

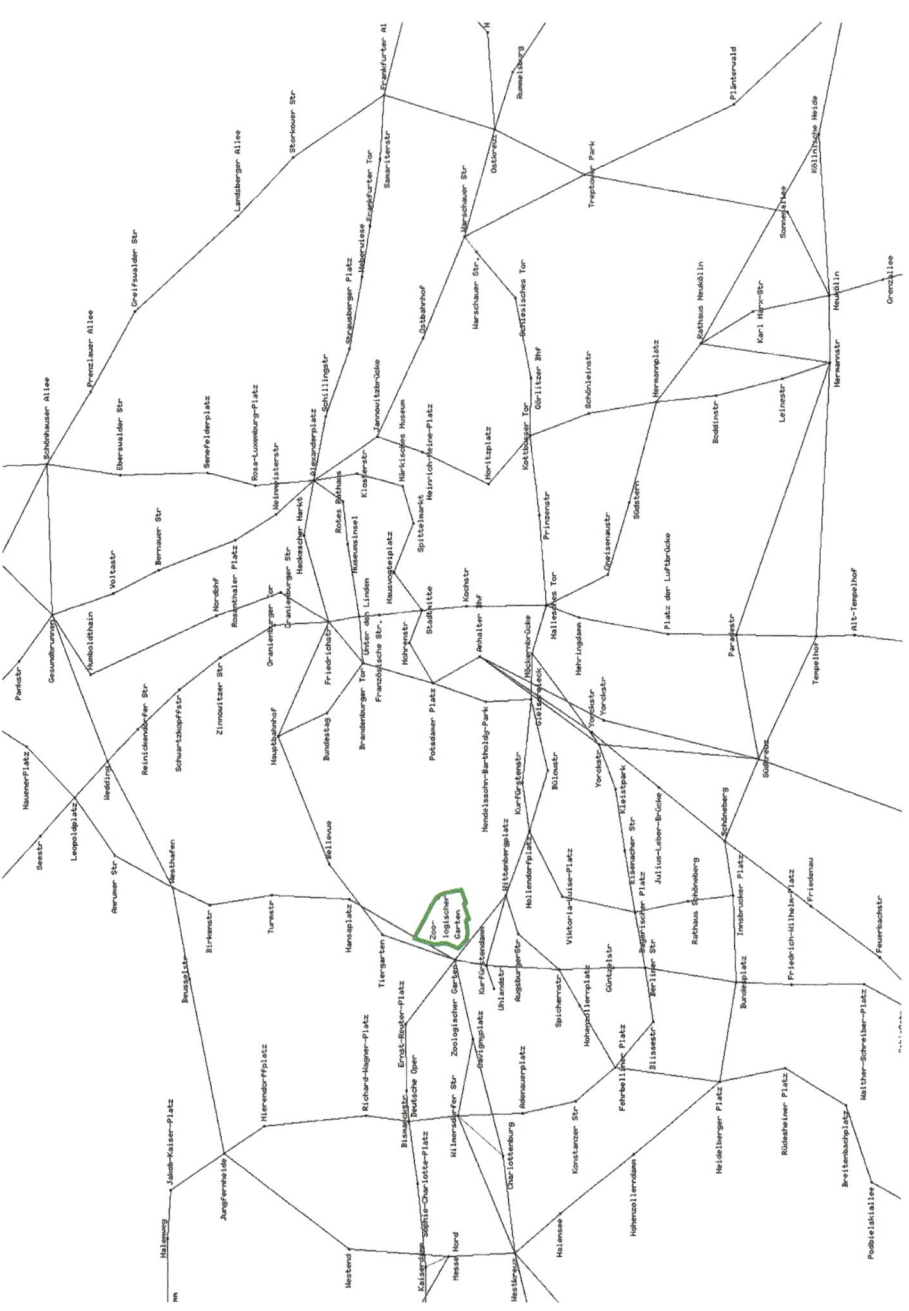

Fig. 16.2 Extract from the U- and S-Bahn-Net in Berlin

16.6 Construction

```
type Station interface {

  Object

  Set (l Line, nr uint, n string, b byte, y, x float64)
  Line() Line
  Number() uint
  Pos() (float64, float64)
  Umstieg()
  Renumber (l Line, nr uint)
  Equiv (Y any) bool
  EditScale()
  UnderMouse() bool
  Write (b bool)
```

The Connections

Also connections have the type Object—the prerequisite for them to be inserted as *edges* in graphs (see Sect. 3.5.11).

```
package track
import (. "µU/obj"; "bus/line")

type Track interface { // connection with line and natural number
                       // as value (medium travel time in minutes)
  Object
  Valuator

// x belongs to the line l and haa the value f.
  Def (l line.Line, f uint)

// (x, y) and (x1, y1) are the positions of the endpoints
// of the calling track
  SetPos (x, y, x1, y1 float64)

// x is written to the screen, for b == true
// in the colour of its line, otherwise in black.
  Write (b bool)
}
```

The Lines

The specification of the lines consists of the enumeration of the U- and S-Bahn lines in Berlin and the colours assigned to them by the BVG.

```
package line
import "µU/col"

type Line byte; const (
  Footpath = Line(iota)
  U1; U2; U3; U4; U5; U6; U7; U8; U9; S1; S2; S25; S26; S3
  S41; S45; S46; S47; S5; S7; S75; S8; S85; S9; Zoo; BG; NLines)
var (
  Text = []string {"F", "U1", "U2", "U3", "U4", "U5", "U6",
                   "U7", "U8", "U9", "S1", "S2", "S25", "S26",
                   "S3", "S41", "S45", "S46", "S47", "S5",
                   "S7", "S75", "S8", "S85", "S9", "Zoo", "BG"}
  Colour = []col.Colour {col.White (),
                         col.New3n ("U1", 85, 184, 49),
                         col.New3n ("U2", 241, 71, 28),
                         ...
```

```
                            col.New3n ("S1", 119, 95, 176),
                            col.New3n ("S2",  19, 133,  75),
                            ...
)
```

16.6.2 Implementation

We only show the representations of the data types `stat` and `track` and a section from the file `construct.go` from the package `net`.

The Stations
The representation of `station` is as follows:
```
package stat
import (. "µU/obj"; "µU/time"; "µU/linewd"; "µU/str"
          "µU/col"; "µU/scr"; "µU/scale"; "bus/line")
const (
  dB =  67.62 // km per latitude at 52,5 degrees latitude
  dL = 111.13 // km per longitude
)
type station struct {
              bg, lg float64 // position (latitude and longitude)
              line line.Line
              uint // internal number
          umstieg bool
                  string "name"
      beschriftung byte // positioning of the name on the graphic
}
```

The Connections
Here is the representation of `track`:
```
package track
import (. "µU/obj"; "µU/linewd"; "µU/col"
          "µU/scr"; "µU/scale"; "bus/line")
const dB, dL = 67.62, 111.13 // km per latitude resp. longitude
                             // at 52.5 degrees latitude
type track struct {
              line.Line
    x, y, x1, y2 float64 // positions of the stations
              uint "travel time"
}
```

The Network
The network is represented as a graph. In its construction in the file `net/construct.go`, each station is inserted into this graph in a line of code and connected to the station from the line of code before. We show a short exemplary section from this construction.
```
package net
import ("µU/ker"; "µU/str"; "µU/errh"; . "bus/line"; "bus/stat")

// Actual corner is (1, nr), postactual corner is that one
```

16.6 Construction

```
// that was previously actual.
func ins (l Line, nr uint, k string, b byte, y, x float64) {
  k = str.Lat1 (k)
  station.Set (l, nr, k, b, y, x)
  lastX, lastY = x, y
  netgraph.Ins (station)
}

// Actual corner is (l, nr), postactual corner is that one
// that was previously actual. t is the medium travel time
// from the station in the program line before.
func ins1 (l Line, nr uint, n string, b byte, y, x float64, t uint) {
  x0, y0 := lastX, lastY
  ins (l, nr, n, b, y, x)
  trk.Def (l, t)
  trk.SetPos (x0, y0, x, y)
  netgraph.Edge (trk)
}

func constructNet() {
  ins  (U1, 10, "Uhlandstr___",    U, 52.5030, 13.3276)
  ins1 (U1, 11, " Kurfürstendamm", O, 52.5038, 13.3314, 1)
  ins1 (U1, 12, "Wittenbergplatz", R, 52.5018, 13.3430, 2)
  ins1 (U1, 13, "Nollendorfplatz", L, 52.4994, 13.3535, 2)
  ins1 (U1, 14, "Kurfürstenstr",   O, 52.5001, 13.3615, 1)
  ...
}
```

Index

A
Abstract data object, 9, 22
Abstract data type, 3, 9, 26
Abstract variable, 27
Address directory, 225
Address operator, 26
Alphabet, 238
Appointment, 2, 140
Atom, 222, 227
AVL tree, 93

B
B-tree, 99
Bahnhausen, 279
Bahnheim, 278
Bahnstadt, 279
Block, 2, 126, 269
Board, 189
Branch, 248
Buffer, 78
Byte, 39
Byte sequence, 30, 39

C
Case distinction, 201
Cell, 275, 276
Chalk, 190
Circle, 104, 192, 196, 301
Clear, 39
Clearer, 39
Codelen, 40
Collection, 69
Command, 64
Commands, 60

Complexity, 83
Component, 7–9, 11, 19
Components, 2, 126
Computability, 2, 166
Concrete data type, 23
Conic section, 301
Constants, 12, 275
Constructor, 21, 27, 35
Copy, 37
Cursor, 64
Curve, 192, 195

D
Data object, 11
Data set, 221, 222
Data storage, 202
Decomposition, 9, 13, 19
Depth-first search, 2, 126
Dereferencing operator, 26
Derivation, 238
Direction, 275
Double rotation, 86
Driving order, 269, 281

E
Edit, 64
Effect description, 11
Eisenhausen, 280
Eisenheim, 279
Eisenstadt, 280
Ellipse, 192, 196, 301
Empty, 38
Encode, 40
Entry track, 274

Environment variable, 18
Equal, 29, 37
Equaler, 37
Equality predicate, 25
Error message, 65
Euler path, 104
Event loop, 13
Exit track, 274

F
Fibonacci tree, 92
Field, 63
File, 44
Fixed width font, 32
Flower, 256
Font, 32, 46
Framebuffer, 33
Func main, 22
Function, 9, 173

G
Game of Life, 1, 157
GL, 32
GLUT, 32
Go, 17
Graph, 103, 305

H
Handle, 97
Heap, 81
Herb, 249
Hilbert curve, 245
Hint, 65
Hyperbola, 302

I
Iff, 37
Image, 193, 196
Implementation, 5–7, 14, 123
Import, 20
Incline, 251
Index, 99, 227
Indexer, 99
Inferno program, 225, 228
Instruction, 204
Instructions, 3, 167

Interface, 20, 26
IT system, 2, 120
Iteration, 2, 126

J
Java, 36, 41
Journey, 273

K
Key, 99
Keyboard, 10, 59
Keyword, 3, 141
Kilo, 275
Koch Island, 240
Kraut, 251

L
L-systems, 239
Language, 238
Leonardo number, 94
Letter, 238
LIFO, 78
Line segment, 192
Linux, 32
Loop, 2, 126, 201

M
MacOS, 34
Make, 40
Mask, 226
Memory space, 24, 26, 29
Menue control, 68
Method, 1, 125
Molecule, 222, 227
Mouse, 10, 59

N
Name, 97
Network, 274

O
Obj, 36, 41
Object, 8, 10, 36, 41
Object-oriented, 21

Index

Object-package, 36
Opaque, 27
OpenGL, 32, 265
Order predicate, 25
Ordering relation, 38

P
Package, 19, 35
Parabola, 302
Peano curve, 245
Persistent, 97
Persistent index set, 99
Point sequence, 191
Pointer, 25, 26
Polygon, 192, 195, 251
Power series, 96
Precondition, 1, 125
Predator-prey system, 1, 157
Predicate, 73
Primitive recursion, 15, 179
Principle of secrecy, 14
Print, 66
Priority queue, 80
Processor, 202
Production rule, 238
Program, 4, 168, 203
Program line, 203
Program storage, 202
Push-button interlocking, 269

Q
Queue, 78, 80

R
Rectangle, 192, 196
Recursion, 2, 15, 126, 179, 201
Recursive function, 16, 180
Reference, 26, 27, 30, 74
Reference semantics, 28
Register, 2, 4, 166, 168, 202
Ring list, 75
Road, 270, 274
Robi language, 3, 127
Robilanguage, 1, 125
Robot, 2, 126
Route, 269

S
Screen, 10, 44, 48
Screen mode, 46
Screen package, 48
Selection menu, 13
Sequence, 74, 201
Set, 83
Shortest connection, 104
Shunting movement, 269
Signal, 273, 277
Slice, 39
Software life cycle, 3
Space-filling curve, 244
Specification, 5–7, 20, 21, 123
Ssh, 33
Stack, 78
Stack storage, 202
Start symbol, 238
State, 239, 264, 273, 277
Station, 269, 274
Straight line, 196
Stream, 39
Strings, 43
Structure, 222
Structured programming, 1, 125
Switch, 273
Symbol, 238
Synchronization, 274
System analysis, 3, 121
System architecture, 5, 8, 10, 123

T
m , 66
Teaching project, 2, 120
Test, 12
Three-dimensional bush, 253
Three-dimensional Hilbert Curve, 253
Three-dimensional tree, 256
Tilt, 252
Track, 270
Turing machine, 1, 2, 165, 166
Type size, 24, 27, 29

U
Unix, 44
Usage prerequisite, 11
User interface, 10

User manual, 5, 123

V
Value, 43
Value assignment, 25, 28
Value parameter, 25, 26
Value semantics, 28, 30
Variable, 2, 24, 26, 166
Variables, 12
Visibility, 20

W
WHILE-computability, 181
Windows, 34
With reference semantics, 30
Word, 238
World, 2, 126

X
X-Window, 32, 50

GPSR Compliance

The European Union's (EU) General Product Safety Regulation (GPSR) is a set of rules that requires consumer products to be safe and our obligations to ensure this.

If you have any concerns about our products, you can contact us on ProductSafety@springernature.com

In case Publisher is established outside the EU, the EU authorized representative is:

Springer Nature Customer Service Center GmbH
Europaplatz 3
69115 Heidelberg, Germany

Batch number: 08690097

Printed by Printforce, the Netherlands